Using the Project Management Maturity Model

Using the Project Management Maturity Model

Strategic Planning for Project Management

Third Edition

HAROLD KERZNER, PH.D.

WILEY

Library of Congress Cataloging-in-Publication Data:

Names: Kerzner, Harold, author.
Title: Using the project management maturity model : strategic planning for
  project management / Harold Kerzner, Ph.D.
Description: Third edition. | Hoboken, New Jersey : John Wiley & Sons, Inc.,
  [2018] | Includes bibliographical references and index. |
Identifiers: LCCN 2018048264 (print) | LCCN 2018050929 (ebook) | ISBN
  9781119530879 (Adobe PDF) | ISBN 9781119530824 (ePub) | ISBN 9781119530824
  (pbk.)
Subjects: LCSH: Project management. | Strategic planning.
Classification: LCC HD69.P75 (ebook) | LCC HD69.P75 K494 2018 (print) | DDC
  658.4/04—dc23
LC record available at https://lccn.loc.gov/2018048264

Printed in the United States of America

V10008208_021919

Contents

Chapter 8   Level 4: Benchmarking   97

Chapter 9   Level 5: Continuous Improvement   109

Chapter 10   Sustainable Competitive Advantage   139

Chapter 11   Advanced Project Management Maturity Assessments   147

Chapter 12   How to Conduct a Project Management Maturity Assessment   173

Chapter 13   Using the PMMM to Extract Best Practices   187

Chapter 14   Case Studies   201

Appendix   The Kerzner Project Management Maturity Model   235

Preface

Excellence in project management cannot occur, at least not within a reasonable time frame, without some form of strategic planning for project management. Although the principles of strategic planning have been known for several decades, an understanding of their applicability to project management has been slow in acceptance. Today, as more companies recognize the benefits that project management can provide to their bottom line, the need for strategic planning for project management has been identified as a high priority.

The definition of project management maturity is constantly changing as the landscape for project management changes. Techniques such as agile and Scrum have forced us to rethink our definitions of project management maturity. Maturity in project management is a continuously evolving process. Traditional project management maturity models must now allow for customization because each company can have a different definition of project management maturity. One size no longer fits all.

This book is broken down into three major parts. The first part, Chapters 1 to 3, discusses the principles of strategic planning and how it relates to project management, the definition of project management maturity, and the need for customization. The second part, Chapters 4 to 9, details the project management maturity model (PMMM), which will provide organizations with general guidance on how to perform strategic planning for project management. The various levels, or stages of development, for achieving project management maturity, and the accompanying assessment instruments, can be used to validate how far along the maturity curve the organization has progressed. The PMMM has been industry validated. One large company requires that, each month, managers and executives take the assessment instruments and then verify that progress toward maturity is taking place from reporting period to reporting period. Other companies have used PMMM to assess the corporation's knowledge level regarding project management as well as a means for assessing the needs for a project management office, a best practices library, external and internal benchmarking, and the identification of the type of project management training needed. Options exist for customization in the various levels.

Chapters 10 to 13 discuss some relatively new concepts in project management such as how assessments can be made to measure the firm's growth using PM 2.0 and

PM 3.0. Many of these concepts are the result of strategic planning for project management activities.

Perhaps the major benefit of the PMMM is that the assessment instruments for each level of maturity can be customized for individual companies. This customization opportunity makes *Using the Project Management Maturity Model* highly desirable as a required or reference text for college and university courses that require students to perform an individual or group research project. The book should also be useful as a required text for graduate courses on research methods in project management. In addition, the book can be used as an introduction to research methods for project management benchmarking and continuous improvement, as well as providing a brief overview of how to design a project management methodology.

Seminars on strategic planning for project management using this book, as well as other training programs on various project management subjects, are available by contacting Lori Milhaven, Vice President, at the International Institute for Learning, 212-515-5121. Contact can also be made through the website (iil.com). PowerPoint slides of the material in this book may also be found on the supporting website, www.wiley.com/go/pmmm3e.

Harold Kerzner
International Institute for Learning
110 East 59th Street
New York, NY 10022-1380

Introduction

People often ask me how I came up with the idea for creating a project management maturity model (PMMM). In 1996, the International Institute for Learning (IIL) partnered with Microsoft and Nortel to sponsor a global videoconference where I discussed some of the project management best practices that companies were using. After the broadcast, I was flooded with questions, with conference participants asking me how "quickly" their company could implement some of these best practices and become good at project management. I responded to the participants that maturity and excellence in project management cannot be achieved easily or quickly without some type of strategic direction focusing on project management maturity. The direction soon became the PMMM.

In 1997, when I first prepared the foundation for the PMMM, there were very few maturity models in the marketplace. Today, there are more than 30. Every model has its pros and cons. Some models take a great deal of time to do the assessments, whereas others are fairly quick and cost-effective to use. Some models are more applicable to specific industries, such as construction or IT, whereas other models are more generic.

The PMMM was created to prepare companies for the future rather than the present. To understand this, you must first recognize what makes project management work well. Having an enterprise project management methodology does not necessarily lead to maturity. Having policies and procedures embedded throughout the methodology is also no guarantee that maturity will be forthcoming. Even following the *PMBOK®
Guide* exactly cannot guarantee maturity.

Before you start sending me nasty e-mails, let me state my position on the previous paragraph. Project management methodologies based on rather rigid policies and procedures were created because management wanted standardization in the way that projects were planned, scheduled, and controlled. This was a necessity because executives had concerns about the ability of their project managers to make the correct decisions. Some people have argued that these rigid approaches mandated "obedience to regulations" and limited the freedom that most project managers need. The problem with standardization is that it often pulls people out of their comfort zones, and they

---

PMBOK is a registered mark of the Project Management Institute, Inc.

must work differently when assigned to projects. People who are asked to work outside of their comfort zone often dislike working on project teams and may look forward to the end of the project so they can return to their previous assignment. What I have observed in the past five decades is that project management excellence comes from four critical components:

- Effective communications
- Effective cooperation
- Effective teamwork
- Trust

With this in mind, the PMMM is significantly more behavioral than quantitative. People manage projects; methodologies function as supporting tools. You can have the greatest methodology in the world and still not reach a level of maturity, because the correct human behavior is not in place. Maturity in project management occurs when people work together correctly. The PMMM assessments focus on people interacting with other people rather than just tools.

Over the years, executives have seen the benefits of using project management correctly. As executives demonstrate more trust in project managers' capabilities, rigid methodologies are being replaced with forms, guidelines, templates, and checklists. Today, at the beginning of a project, the project manager will walk through the "cafeteria" and select from the shelves only those forms, guidelines, templates, and checklists that are appropriate for that project and that client. We now have flexible methodologies, or frameworks. If the project manager believes that this project is a very low risk, then the project manager may not want to follow or even use the "Risk Management" section of the *PMBOK® Guide*. Project managers are now being given more freedom over how to apply project management practices to satisfy the customer's needs. This leads to customer satisfaction and repeat business.

But even with this new freedom, project managers must still recognize the importance of the behavioral assessments in the PMMM, which focus on effective communication, cooperation, teamwork, and trust. Behavioral assessments indicate whether people believe that they are working within their comfort zone. If continuous improvements are made correctly (i.e., Level 5 of the PMMM) and people are happy with their comfort zone, some degree of project management maturity can be achieved quickly. The focus in the PMMM is that people manage projects; people manage tools; tools by themselves manage neither people nor projects. As a former Air Force lieutenant general stated, "You must never allow the tool to control the hand that's holding it." Maturity models should certainly include an assessment of whether the organization has the right tools and practices in place. But in my opinion, there should be an equal or possibly heavier emphasis on the necessary human behavior.

A few years ago, I was interviewed for an article on maturity models with an emphasis on the PMMM. Following are some of the questions I was asked.

Q1: How much project management maturity does a company really need?

The amount of maturity a company needs is quite often customer driven rather than internally driven. Whenever a contractor allows its customer to become more mature than it is, very unfavorable results can occur. Among them, (1) the customer tells the contractor how the work should be done, (2) the customer may perform the work by themselves, and (3) the customer may seek out a more mature contractor during competitive bidding. Therefore, companies that rely heavily on external customers for their revenue stream, such as project-driven companies, must *never* allow their customers to achieve a greater degree of maturity than theirs. For these companies, project management maturity is a necessity for survival, and the frequent use of a maturity model should be mandatory.

Companies today should be willing to perform a frequent self-assessment to make sure that the firm is continuously improving and reaching some level of maturity. During competitive bidding activities, customers are now asking contractors to show how mature their organization is with regard to project management. Maturity assessments could be the difference between winning and losing a potential contract.

Q2: Which industries are making the greatest strides toward maturity?

First of all, it is questionable if maturity can ever be accurately defined or measured because saying that you are mature in project management might imply that there is no further room for improvement. This can lead to complacency and a loss of competitiveness. But if I were asked which industries appear to be more mature than others, I would begin with project-driven companies that rely on competitive bidding for their revenue stream and must sell their delivery system as well as the expected project outcomes. They have come to the realization that they must try to remain more mature in project management than their customers simply to stay in business. As companies become more mature, tremendous pressure is exerted on their supplier base to improve in project management, and organizational maturity assessment information is appearing as a requirement in the RFP. In fact, reaching certain maturity levels in project management has now become a competitive weapon during competitive bidding activities. In general, organizations where projects have profit expectations and the project manager is responsible for generating the profits appear to mature faster than organizations where there are no profit margins assigned to projects.

Q3: What can companies just starting out on the maturity process learn from those leading the pack?

It is always better to learn from the mistakes of others rather than from your own mistakes. Several lessons can be learned. First, strategic planning for project management maturity is essential, even a necessity. Without guidance from some sort of project management maturity (strategic planning) model, achieving maturity could take decades as you learn from your own mistakes. All project management maturity models, in my opinion, are a form of strategic planning. Second, there must be a corporate commitment (especially at the executive levels) for maturity to occur, and the executives must see the value in achieving a defined level of maturity in a reasonable time frame. There are assessment questions on this in the PMMM. Third, there must exist a dedicated organization that drives the maturity process, and this normally becomes the responsibility of the project management office (PMO). Companies where PMOs take the lead in the assessment and continuous improvement processes generally reach levels of maturity more quickly than those that do not have any involvement by the PMO.

Q4: Should every company be pursuing maturity? What keeps some companies back?

Given the fact that many executives today view their company as being a stream of projects, the project management approach permeates the entire organization, mandating that maturity is necessary. Only those companies that want to stay in business and remain competitive should pursue maturity. The alternative is rather unpleasant.

Pursuing and even obtaining some degree of maturity does not guarantee that business will improve. The company must still make realistic and practical business decisions, and executives must visibility promote the continuation of project management excellence.

Q5: What would prevent a company from achieving the height of maturity?

Other than the behavioral issues I discussed before, several factors prevent companies from achieving maturity. These include: (1) executives not seeing the value in project management or in project management maturity; (2) executives not recognizing that project management maturity is now a competitive weapon; (3) executives not realizing the importance of project management maturity to customers and competitors; (4) executives not willing to establish a PMO to guide the maturity process; and (5) executives not willing to commit sufficient resources to achieving maturity. Obviously, there is a common theme in these five factors: *executives*. Hence, executive education has been a priority in recent years. Executives must see the return on investment as a result of using assessment instruments such as the PMMM. There are assessment questions on executive expectations and involvement in the PMMM. And once again, this emphasizes the importance of behavioral assessment.

Q6: There are a number of available maturity models in the marketplace. How does a company choose the maturity model that's right for its needs?

There are several PM maturity models in the marketplace. And while they all have a different approach, they all have the same ultimate objective: maturity! The decision of which model is best for a given company might be based on the time frame allotted, number of resources available for implementing changes that are needed, pressure from customers, maturity level of competitors, and whether the company is project- or non-project-driven.

Today, there are numerous papers published as well as master's degree and Ph.D. theses that benchmark the various models. Even though I am somewhat partial to the PMMM, there are other models for maturity assessments that are equally as good or better for certain applications. What should be important is not necessarily what model you select but the fact that you are doing an assessment. In my opinion, all of the models in the marketplace provide some type of value if used properly.

Q7: What components differentiate the best models from the pack?

I think that two primary components must be considered: simplicity and assessment capability. Published articles on maturity model benchmarking may have dozens of components, many of which are industry specific. I prefer just these two components as starters. The prospect of using a complex maturity model may very likely scare away senior management because they may not be able to determine time frames or resources needed to achieve maturity. With maturity models, complexity breeds avoidance. With regard to capability, assessment instruments are needed to identify areas of improvement and show that progress is being made and that continuous improvements in project management are adding value to the business.

Q8: What are the advantages and disadvantages of adopting a model using levels of maturity versus one that does not?

Using a maturity model without levels is like managing a five-year project without life-cycle phases. There is often a lack of structure and discipline, possibly a lack of metrics, and no well-established decision points for corrective action. I certainly would not like to manage a project without these elements in place.

Q9: How can a company maintain momentum after reaching a plateau?

The answer to this question is simple: executive support, executive support, and executive support. Need I say more?

Q10: How long and how much money does it typically take companies to reach the higher levels of maturity? Is it worth the investment in time and money? Can you prove it?

It has been my experience that the single most important force for achieving higher levels of maturity (other than continuous executive support) is the early-on establishment of a PMO. The PMO becomes the major driver for the maturity process. Without a PMO, it may take three to five years to reach certain initial levels of maturity. With early establishment of a PMO, however, and the right people assigned to the PMO, it may take only two years or less. The problem with deciding upon a time frame for maturity is heavily based on someone's definition of maturity, the speed with which tools are either purchased or developed, and the commitment to the right levels of project management education. Any organization can develop all the tools necessary to achieve maturity. But if the organization does not understand the benefits and value of project management or the use of the tools, what has it really accomplished? Maturity *is not* just the development of tools or processes. Maturity is the effective *use* of these instruments, and continuous improvement in the use of these instruments using captured best practices. Whenever companies ask me whether the investment of time and money to obtain maturity is worth it, my response is simple. You know the amount of money needed to achieve a certain level of maturity. But what is the cost or opportunity loss of not achieving it? Is it possible for the opportunity loss to be at least an order of magnitude greater than the cost of achieving maturity? You bet!

CHAPTER ONE

The Need for Strategic Planning for Project Management

► Introduction

For more than 50 years, American companies have been using the principles of project management to get work accomplished. Yet, for more than 40 of these years, very few attempts were made to recognize project management as a core competency for the company. There were three reasons for this resistance to project management. First, project management was initially viewed as simply a scheduling tool for the workers. Second, since this scheduling tool was thought to belong at the worker level, executives saw no reason to look more closely at project management, and thus failed to recognize the true benefits it could bring. Third, executives were fearful that project management, if viewed as a core competency, would require them to decentralize authority, to delegate decision-making to the project managers, and thus to diminish the executives' power and authority base.

► Misconceptions

As the twenty-first century approached, project management began to mature in virtually all types of organizations, including those firms that were project-driven, those that were non–project-driven, and hybrids. Knowledge concerning the benefits project management offered now permeated all levels of management. Project management came to be recognized as a process that would increase shareholder value.

This new knowledge regarding the benefits of project management allowed us to dispel the illusions and misconceptions that we had believed in for over 40 years. These misconceptions or past views are detailed next, together with current views.

Cost of Project Management

- *Misconception:* Project management will require more people and increase our overhead costs.

- *Present view:* Project management allows us to lower our cost of operations by accomplishing more work in less time and with fewer resources, without any sacrifice in quality or value.

Profitability

- *Misconception:* Profitability may decrease.
- *Present view:* Profitability will increase.

Scope Changes

- *Misconception:* Project management will increase the number of scope changes on projects, perhaps due to the project manager's desire for extreme creativity.

- *Present view:* Project management provides us with better control of scope changes. Good project managers try to avoid unnecessary scope changes.

Organizational Performance

- *Misconception:* Because of multiple-boss reporting, project management will create organizational instability and increase the potential for conflicts.

- *Present view:* Project management makes the organization more efficient and effective through better application of organizational behavior principles.

Customer Contact

- *Misconception:* Project management is really "eyewash" for the customer's benefit.

- *Present view:* Project management allows us to develop a closer working relationship with our customers. This can lead to increased business opportunities.

Problems

- *Misconception:* Project management will end up creating more problems than usual.

- *Present view:* Project management provides us with a structured process for effectively solving problems.

Applicability

- *Misconception:* Project management is applicable only to large, long-term projects such as in the aerospace, defense, and construction industries.

- *Present view:* Virtually all projects in all industries can benefit from the principles of project management.

Quality

- *Misconception:* Project management will increase the potential for quality problems.
- *Present view:* Project management will increase the quality and value of our products and services.

Power/Authority

- *Misconception:* Multiple-boss reporting will increase problems related to power and authority.
- *Present view:* Project management will reduce power/authority problems.

Focus

- *Misconception:* Project management focuses on suboptimization by looking at the project only.
- *Present view:* Project management allows us to make better decisions for the best interest of the company.

Project's End Result

- *Misconception:* Project management delivers products to a customer.
- *Present view:* Project management delivers business solutions to a customer.

Competitiveness

- *Misconception:* The cost of project management may make us noncompetitive.
- *Present view:* Project management will increase our business (and even enhance our reputation).

▶ Project Management Becomes a Strategic Competency

As senior management became more knowledgeable about project management, the misconceptions subsided and appreciation and understanding of how project management could benefit the organization grew. Today's view of project management includes the following:

- Project managers should no longer consider themselves as simply managing a project. Instead, they should see themselves as managing part of a business.
- Project managers are now expected to make both project- and business-related decisions, whereas previously most business-related decisions were made by the project sponsor or governance committee.
- Project managers are now managing both strategic as well as tactical projects. Previously, strategic activities were assigned to line managers rather than project managers.
- Project management is now seen as the delivery system for achieving strategic business objectives.

- Project management produces deliverables and outcomes that can be converted into business benefits and business value.

- Companies that wish to prepare for the future perform a study every year or two to determine which four or five career-path positions are an absolute necessity for the company to survive. Project management often makes the list and is now regarded as a strategic competency rather than just another career-path position.

- As a strategic competency, project managers are expected to have a better understanding than their predecessors had concerning the business itself and strategic planning.

▶ General Strategic Planning

Strategic planning is the process of formulating and implementing decisions about an organization's future direction. This is shown in Figure 1.1. It is vital to every organization's survival because it is the process by which the organization adapts to its ever-changing environment and achieves its strategic objectives. The process is applicable to all management levels and all types of organizations.

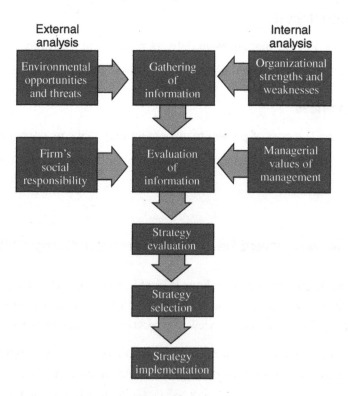

Figure 1.1 Basic strategic planning.

The critical box in Figure 1.1 is the last one, Strategy implementation. People tend to focus heavily on the steps to get to strategy formulation and fail to realize that project management is the delivery system necessary to implement the strategy.

As an example, a Fortune 500 company hired a consulting company to analyze all the firm's product lines and to provide the firm with advice on business strategy. For a week, the executives met with the consultants. The beginning of the following week, after the consultants left, the executives convened in the board room to review what they had learned. The conclusion was that the consultants told them "what to do" but not "how to do it." The executives realized quickly that project management would be needed to convert the "what" to "how." The Human Resources department was given the mandate to begin training in project management so that the firm could become reasonably mature in delivering strategic objectives and to perform periodic assessments to see that progress was being made.

In another example, the Industrial Products Group (IPG) of a Fortune 500 company recognized quickly the need for project management to help achieve strategic business objectives. Part of the company's business was an Aerospace Group that appeared to be reasonably mature in project management because it had been working on government contracts for more than 20 years. Several managers from the Aerospace Group were permanently transferred into the IPG in hopes of accelerating project management maturity.

After a short while, assessments were conducted that showed progress was not being made and, in some situations, conditions had gotten worse. The IPG then realized that many of the tools and processes used in the Aerospace Group were either too complex or not appropriate for the IPG. The company learned that the tools, forms, guidelines, templates, and checklists that helped bring some level of maturity in one division may not bring the same level of maturity in another division. Customization would be required.

▶ Participation by the Project Manager in Strategic Planning

Historically, project managers were brought on board a project after the project was approved, the business case was created, and the priority was set. Then the project manager was told how much money they had and the time frame. Constraints were often established by senior management or marketing/sales with no input by the project manager. Then the project manager was expected to meet unrealistic expectations.

As stated previously, today's project managers are more actively involved in business decisions and responsible for achieving business objectives. As such, they are being brought on board earlier and in some companies are participating in strategic planning activities.[1] The formulation process shown in Figure 1.1 is the high-level process of deciding where you want to go, what decisions must be made, and when they must be made to get there in a timely manner. It is the process of defining and understanding the business you are in and how to remain competitive within that business. The outcome of successful formulation results in the organization doing the right thing

---

[1] Traditional life-cycle phases are now being replaced by investment life-cycle phases where the project managers are brought on board earlier than before. This is a necessity so that the project manager understands the business benefits and value expected from this project or the portfolio of projects. For additional information on this, see Harold Kerzner, *Project Management Best Practices: Achieving Global Excellence* (Hoboken: John Wiley, 2018), 715–737.

in the right way (i.e. it results in project management) by producing goods or services for which there is a demand or need in the external or internal environment. When this occurs, we say the organization has been effective as measured by market response, such as sales and market shares or customer acceptance. A good project management methodology, whether a rigid methodology or a flexible methodology such as agile or Scrum, can lead to better customer satisfaction and a greater likelihood of repeat business. All organizations must be effective and responsive to their environments to survive in the long run.

All too often, projects are selected and approved without an accurate understanding of the organization's capabilities at that time. This occurs because executive management does not know how much additional work they can undertake without overburdening the existing labor force. The benefit of having project managers brought on board this early is that they can provide information related to the following questions:

- How many resources will be needed?
- What skill levels must the resources possess?
- Does the organization currently have sufficient resources available internally?
- Will the resources be assigned full-time or part-time?
- Can this project be accomplished with a virtual project team?

The formulation process is performed at the top levels of the organization, but involvement by the project manager can accelerate downstream decision-making and possibly reduce the number of action items. Here, top management values provide the ultimate decision template for directing the course of the firm.

Formulation:

- Scans the external environment and industry environment for changing conditions.
- Interprets the changing environment and the enterprise environmental factors in terms of opportunities or threats.
- Analyzes the firm's resource base for asset strengths and weaknesses.
- Defines the mission of the business by matching environmental opportunities and threats with resource strengths and weaknesses.
- Sets goals for pursuing the mission based on top management values and sense of responsibility.

The second step in strategic planning, implementation, translates the formulated plan into a reality. At this point, project management involvement should be mandatory. Implementation involves all levels of management in moving the organization toward its mission. The process seeks to create a fit between the organization's formulated goal and its ongoing activities or projects. Because implementation involves all levels of the organization, it results in the integration of all aspects of the firm's functioning.

Integration management is a vital core competency of project management. As shown in Figure 1.2, there is a hierarchy of plans, and they all require integration both within and across strategic business units (SBUs). Project management is now recognized as a vehicle for the integration of just about any type of plan for any type of project.

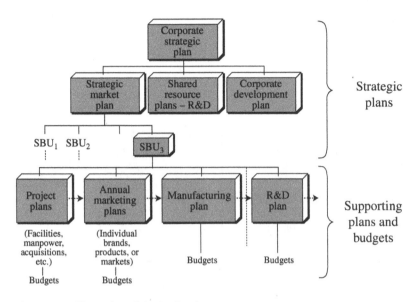

Figure 1.2   Hierarchy of strategic plans.

Middle- and lower-level managers spend most of their time on implementation activities. Effective implementation, supported by a mature project management organization, results in stated objectives, action plans, timetables, policies and procedures, and in the organization moving efficiently toward fulfillment of its mission.

▶ What Is Strategic Planning for Project Management?

Strategic planning for project management is the *development of the necessary tools for project management*. Some companies have as many as 50 tools that the project manager can use. The tools, when combined, form a methodology or framework that can be used over and over again and that will produce a high likelihood of achieving the project's objectives. Although strategic planning for the methodology and execution of the methodology or framework does not guarantee profits or success, it does improve the chances of success.

One primary advantage of developing a flexible or inflexible methodology is that it provides the organization with a consistency of action. As the number of interrelated functional units in organizations has increased, so have the benefits from the integrating direction afforded by the project management implementation process.

Methodologies need not be complex. Figure 1.3 shows the "skeleton" for the development of a simple project management methodology. The methodology begins with a project definition process, which is broken down into a technical baseline, a functional or management baseline, and a financial baseline. The technical baseline includes, at a minimum:

■ Statement of work (SOW)

■ Specifications

■ Work breakdown structure (WBS)

■ Timing (i.e. schedules)

■ Spending curve (S curve)

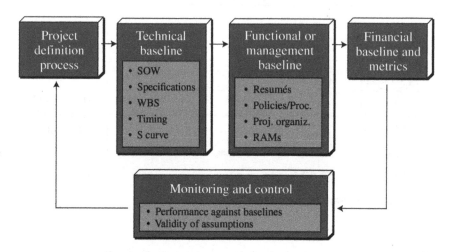

Figure 1.3 Methodology structuring.

The functional or management baseline indicates how you will manage the technical baseline. This includes:

■ Résumés of the key players, if needed

■ Project policies and procedures

■ The organization for the project team

■ Responsibility assignment matrices (RAMs)

The financial baseline identifies how costs will be collected and analyzed, how variances will be explained, and how reports will be prepared. Altogether, this process can be applied to every project.

Another advantage of strategic project planning is that it provides a vehicle for the communication of progress in accomplishing the overall goals and objectives to all levels of management in the organization. It affords the potential for a vertical feedback loop from top to bottom, bottom to top, and functional unit to functional unit. The process of communication and its resultant understanding helps reduce resistance to change. It is extremely difficult to achieve commitment to change when employees do not understand its purpose. The strategic project planning process gives all levels an opportunity to participate, thus reducing the fear of the unknown and possibly eliminating resistance.

The final and perhaps the most important advantage is the thinking process required. Planning is a rational, logically ordered function. This is what a structured methodology or framework provides. Many managers caught up in the day-to-day action of operations will appreciate the order afforded by a logical thinking process. Methodologies can be based on sound, logical decisions and customized for a client. Figure 1.4 shows the logical decision-making process that could be part of the project-selection process for an organization. Checklists can be developed for each section of Figure 1.4 to simplify the process.

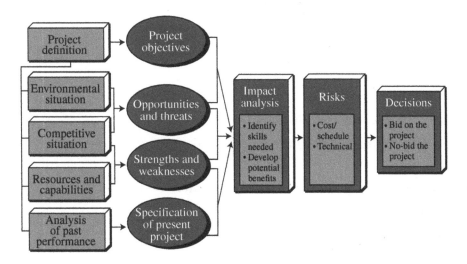

Figure 1.4 Project-selection process.

The first box in Figure 1.4 is the project-definition process. At this point, the project-definition process simply involves a clear understanding of the objectives, which should be defined in both business and technical terms. Based on the type of project, the definition of the project may evolve as the project progresses.

The second box is an analysis of the environmental situation, which is similar to the enterprise environmental factors but with a greater understanding of the business base. This includes a market feasibility analysis to determine:

- The potential size of the market for the product
- The potential risks of product liability
- The capital requirements for the product
- The market position on price
- The expected competitive response
- The regulatory climate, if applicable
- The degree of social acceptance
- Human factors (e.g. unionization)

The third box in Figure 1.4 is an analysis of the competitive situation and includes:

- The overall competitive advantage of the product
- Opportunities for technical superiority:
  - Product performance
  - Patent protection
  - Exceptional price-quality-value relationship
- Business attractiveness:
  - Type and nature of competitors
  - Structure of the competition/industry
  - Differences among competitors (price, quality, etc.)
  - Threat of substitute products
- Competitive positioning:
  - Market share
  - Rate of change in market share
  - Perceived differentiation among competitors and across various market segments
  - Positioning of the product within the product line
- Opportunities for market positioning:
  - Franchises
  - Reputation/image
  - Superior service
- Supply chain management:
  - Ownership of raw material sources
  - Vertical integration
- Physical plant opportunities:
  - Locations
  - Superior logistics support
- Financial capabilities:
  - Available capital
  - Credit rating impact
  - Wall Street support
- Efficient operations management:
  - Inventory management
  - Production
  - Distribution
  - Logistics support

The next box in Figure 1.4 is resources and capabilities. Analysis of resources and capabilities, combined with the analysis of competitive positioning just discussed, allows you to determine our strengths and weaknesses. Identifying opportunities and threats lets you identify what you *want* to do. However, it is knowing your strengths and weaknesses that lets you identify what you *can* do. Therefore, the design of any type of project management methodology must be based heavily on what the organization can do.

Internal strengths and weaknesses can be defined for each major functional area. The design of a project management methodology can exploit the strengths in each functional area and minimize its weaknesses. Not all functional areas will possess the same strengths and weaknesses.

The following illustrate typical strengths or weaknesses for various functional organizations:

- Research and development:
  - Ability to conduct basic/applied research
  - Ability to maintain state-of-the-art knowledge
  - Technical forecasting ability
  - Well-equipped laboratories
  - Proprietary technical knowledge
  - An innovative and creative environment
  - Offensive R&D capability
  - Defensive R&D capability
  - Ability to optimize cost with performance
- Manufacturing:
  - Efficiency factors
  - Raw material availability and cost
  - Vertical integration abilities
  - Quality assurance system
  - Relationship with unions
  - Learning curve applications
  - Subsystems integration
- Finance and accounting:
  - Cash flow (present and future projections)
  - Forward pricing rates
  - Working capital requirements
- Human resource management:
  - Turnover rate of key personnel
  - Recruitment opportunities

- Promotion opportunities
- Having a project management career path
- Quality of management at all levels
- Public relations policies
- Social consciousness
- Marketing:
  - Price-value analysis
  - Sales-forecasting ability
  - Market share
  - Life-cycle phases of each product
  - Brand loyalty
  - Patent protection
  - Turnover of key personnel

Having analyzed what you can do, you must now look at past performance to see if there are any applicable lessons learned files that could impact the current project or selection of projects. Analysis of past performance, as shown in Figure 1.4, is usually the best guide for the specifications of the present project.

Figure 1.4 represents a rough template of whether or not to undertake a project. This type of decision-making process is critical if you are to improve your chances of success. Historically, less than 10 percent of R&D projects make it through full commercialization where all costs are recovered. Part of that problem has been the lack of a structured approach for decision-making, project approval, and project execution. All this can be satisfied with a sound project management methodology.

In the absence of an explicit project management approach, decisions are made incrementally. A response to the crisis of the moment may result in a choice that is unrelated to, and perhaps inconsistent with, the choice made in the previous moment of crisis. Discontinuous choices serve to keep the organization from moving forward. Contradictory choices are a disservice to the organization and may well be the cause of its demise. Such discontinuous and contradictory choices occur when decisions are made independently to achieve different objectives, even though everyone is supposedly working on the same project. When the implementation process is made explicit, however, objectives, missions, and policies become visible guidelines that produce logically consistent decisions.

Small companies usually have an easier time performing strategic planning for project management excellence. Large companies with highly diversified product lines and multiple management styles find that institutionalizing changes in the way projects are managed can be very complex. Innovation and creativity in project management can be a daunting, but not impossible, task.

► Executive Involvement

Senior management's involvement in strategic planning is essential if the process is to move ahead quickly and if full employee commitment and acceptance is to be achieved. The need for involvement is essential:

- A visible general endorsement is mandatory.
- An executive champion (not necessarily a sponsor) must be assigned.
- The executive champion must initiate the process.
- The executive champion must make sure the ideas/aspirations of senior management are included throughout the methodology.
- The executive champion must verify the validity of the corporate assumptions, including:
  - Forward pricing rate data
  - Targeted customers/industries
  - Reporting requirement for senior management
  - Strategic trends
  - Customer-interfacing requirements

If senior management's support is not visible from the onset, then:

- The workers may believe that senior management is not committed to the process.
- Functional managers may hesitate to provide valuable support, believing that the process is unreal.
- The entire process may lack realism and waste time.

Another critical function of senior management is determining *strategic timing*. A strategic plan is a timed sequence of conditional moves that involve the deployment of resources. The executive champion must either develop or approve the strategic-timing activities, which include:

- Establishing the timetable for major moves
- Establishing resource requirements and ensuring availability
- Providing funding and release time for critical assets and hardware/software purchases to support the project management systems

► Critical Success Factors for Strategic Planning

Critical success factors for strategic planning for project management include those activities that must be performed if the organization is to achieve its long-term objectives. Most businesses have only a handful of critical success factors. However, if even one of them is not executed successfully, the business's competitive position may be threatened.

The critical success factors in achieving project management excellence apply equally to all types of organizations, even those that have not fully implemented their project management systems. Although most organizations are sincere in their efforts to fully implement their systems, stumbling blocks are inevitable and must be overcome. Here's a list of common complaints from project teams:

- There's scope creep in every project and no way to avoid it.

- Completion dates are set before project scope and requirements have been agreed on.

- Detailed project plans identifying all the project's activities, tasks, and subtasks are not available.

- Projects emphasize deadlines. We should emphasize milestones, quality, and benefits and value received, not time.

- Senior managers don't always allow us to use pure project management techniques. Too many of them are still date-driven instead of requirements-driven. Original target dates should be used only for broad planning.

- Outdated project management techniques are still being used on most projects. We need to learn how to manage from a plan and how to use shared resources.

- Sometimes we are pressured to provide low estimates to win a contract, but then we must worry about how we'll accomplish the project's objectives.

- There are times when line personnel not involved in a project change the project budget to maintain their own chargeability. Management does the same.

- Hidden agendas come into play. Instead of concentrating on the project, some people are out to set precedents or score political points.

- We can't run a laboratory without equipment, and equipment maintenance is a problem because there's no funding to pay for the materials and labor.

- Budgets and schedules are not coordinated. Sometimes we spend money according to the schedule but are left with only a small percentage of the project activities complete.

- Juggling schedules on multiple projects is sometimes almost impossible.

- Sometimes we filter information from reports to management because we fear sending them negative messages.

- There's a lot of caving in on budgets and schedules. Trying to please everyone all the time is a trap.

▶ Identifying Strategic Resources

All businesses have corporate competencies and resources that distinguish them from their competitors. These competencies and resources are usually identified in terms of a company's strengths and weaknesses. Deciding what a company *should* do can only be achieved after assessing strengths and weaknesses to determine what the company *can* do. Strengths support windows of opportunities, whereas weaknesses create limitations. What a company can do is based on the quality of its resources.

Strengths and weaknesses can be identified at all levels of management. Senior management may have a clearer picture of the overall company's position in relation to the external environment, whereas middle management may have a better grasp of internal strengths and weaknesses. Unfortunately, most managers do not think in terms of strengths and weaknesses; as a result, they worry more about what they *should* do than about what they *can* do.

Large firms have vast resources with strong technical competency, but they often react slowly when change is needed. Small firms can react quickly but have limited strengths. Any organization's strengths and weaknesses can change over time and must, therefore, be closely monitored.

■ Tangible Resources

In basic project management courses, the strengths and weaknesses of a firm are usually described in the terms of its tangible resources. The most common classifications for tangible resources are:

- Equipment
- Facilities
- Manpower
- Materials
- Money
- Information/technology

Another representation of resources is shown in Figure 1.5. Unfortunately, these crude types of classification do not readily lend themselves to an accurate determination of internal strengths and weaknesses for project management. A more useful classification would be human resources, nonhuman resources, organizational resources, and financial resources.

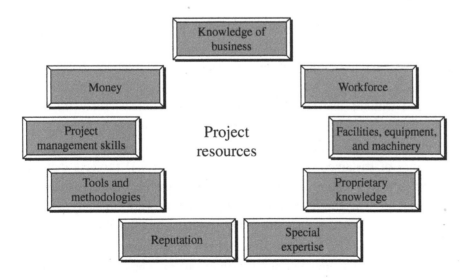

Figure 1.5 Project resources.

Human Resources

Human resources are the knowledge, skills, capabilities, and talent of the firm's employees. This includes the board of directors, managers at all levels, and employees. The board of directors provides the company with considerable experience, political astuteness, and connections, and possibly sources of borrowing power. The board of directors is primarily responsible for selecting the CEO and representing the best interest of the diverse stakeholders as a whole.

Top management is responsible for developing the strategic mission and making sure the strategic mission satisfies the shareholders. All too often, CEOs have singular strengths in only one area of business, such as marketing, finance, technology, or production.

The biggest asset of senior management is its decision-making ability, especially during project planning. Unfortunately, all too often senior management will delegate planning (and the accompanying decision-making process) to staff personnel. This may result in no effective project-planning process within the organization and may lead to continuous replanning efforts.

Another important role of senior management is to define clearly its own managerial values and the firm's social responsibility. A change in senior management could result in an overnight change in the organization's managerial values and its definition of its social responsibility. This could require an immediate update of the firm's project management methodology.

Lower and middle management are responsible for developing and maintaining the core technical competencies of the firm. Every organization maintains a distinct collection of human resources. Middle management must develop some type of cohesive organization such that synergistic effects will follow. The synergistic effects produce the core competencies that lead to sustained competitive advantages and a high probability of successful project execution.

Nonhuman Resources

Nonhuman resources are physical resources that distinguish one organization from another. Physical resources include plant and equipment; distribution networks; proximity of supplies; and availability of raw materials, land, and labor.

Companies with superior nonhuman resources may not have a sustained competitive advantage without also having superior human resources. Likewise, a company with strong human resources may not be able to take advantage of windows of opportunity unless it also has strong physical resources. An Ohio-based company had a 30-year history of sustained competitive advantage on R&D projects that were won through competitive bidding. As times changed, however, senior management saw that the potential for megaprofits now lay in production. Unfortunately, to acquire the resources needed for physical production, the organization diluted some of its technical resources. The firm learned a hard lesson that the management of human resources is not the same as the management of nonhuman resources. The firm also had to reformulate its project management methodology to account for manufacturing operations.

Organizational Resources

Organizational resources are the glue that holds all the other resources together. Organizational resources include the organizational structure, the project office, the formal (and sometimes informal) reporting structure, the planning system, the scheduling system, the control system, and the supporting policies and procedures. Decentralization can create havoc in large firms where each SBU, functional unit, and operating division can have its own policies, procedures, rules, and guidelines. Multiple project management methodologies can cause serious problems if resources are shared between SBUs.

Financial Resources

Financial resources are the firm's borrowing capability, credit lines, credit rating, ability to generate cash, and relationship with investment bankers. Companies with quality credit ratings can borrow money at a lower rate than companies with non-quality ratings. Companies must maintain a proper balance between equity and credit markets when raising funds. A firm with strong, continuous cash flow may be able to fund growth projects out of cash flow rather than through borrowing. This is the usual financial-growth strategy for a small firm.

■ Intangible Resources

Human, physical, organizational, and financial resources are regarded as tangible resources. There are also *intangible resources* that include the organizational culture, reputation, brand name, patents, trademarks, know-how, and relationships with customers and suppliers. Intangible resources do not have the visibility that tangible resources possess, but they can lead to a sustained competitive advantage. When companies develop a *brand name*, it is nurtured through advertising and marketing and is often accompanied by a slogan. Project management methodologies can include paragraphs on how to protect the corporate image or brand name.

■ Social Responsibility

Social responsibility is also an intangible asset, although some consider it both intangible and tangible. *Social responsibility* is the public's expectation that a firm will make decisions that are in the best interest of the public as a whole. Social responsibility can include a broad range of topics from environmental protection to consumer safeguards to consumer honesty and employing the disadvantaged. An image of social responsibility can convert a potential disaster into an advantage.

▶ Why Does Strategic Planning for Project Management Sometimes Fail?

We developed a strong case earlier for the benefits of strategic planning for project management. Knowledge about this process is growing, and new information is being disseminated rapidly. Why, then, does this process often fail? Following are some of the problems that can occur during the strategic planning process. Each of these pitfalls must be considered carefully if the process is to be effective.

- *Lack of CEO endorsement:* Any type of strategic planning process must originate with senior management. They must start the process and signal their own aspirations. A failure by senior management to endorse strategic planning may signal line management that the process is unreal.

- *Failure to reexamine:* Strategic planning for project management is not a one-shot process. It is a dynamic, continuous process of reexamination, feedback, and updating.

- *Being blinded by success:* Simply because a few projects are completed successfully does not mean the methodology is correct, nor does it imply that improvements are not possible. A belief that "you can do no wrong" usually leads to failure.

- *Over-responsiveness to information:* Too many changes in too short a time frame may leave employees with the impression that the methodology is flawed or that its use may not be worth the effort. The issue to be decided here is whether changes should be made continuously or at structured time frames.

- *Failure to educate:* People cannot implement successfully and repetitively a methodology they do not understand. Training and education on the use of the methodology is essential.

- *Failure of organizational acceptance:* Company-wide acceptance of the methodology is essential. This may take time to achieve in large organizations. Strong, visible executive support may be essential for rapid acceptance.

- *Failure to keep the methodology simple:* Simple methodologies based on guidelines are ideal. Unfortunately, as more and more improvements are made, there is a tendency to go from informality using guidelines to formality using policies and procedures.

- *Blaming failures on the methodology:* Project failures are not always the result of poor methodology; the problem may be poor implementation. Unrealistic objectives and poorly defined executive expectations are two common causes of poor implementation. Good methodologies do not guarantee success, but they do imply that the project will be managed correctly.

- *Failure to prioritize:* Serious differences can exist in the importance that different functional areas, such as marketing and manufacturing, assign to strategic project objectives. Figure 1.6 shows three projects and how they are viewed differently by marketing and manufacturing. A common, across-company prioritization system may be necessary.

- *Rapid acquisitions:* Sometimes an organization will purchase another company as part of its long-term strategy for vertical integration. Backward integration occurs when a firm acquires suppliers of components or raw materials to reduce its dependency on outside sources. Forward integration occurs when an organization purchases the forward channels of distribution for its products. In either case, the company's projects will now require more work, and this must be accounted for in the methodology. Changes may occur quickly.

Marketing

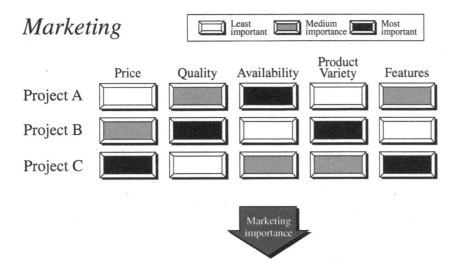

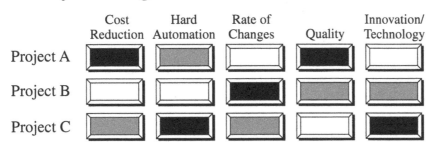

Figure 1.6 Differences in strategic importance.

Only by watching out for these potential problems can a firm hope to avoid them (or at least to minimize their negative effects). This is the path to success in strategic planning for project management.

▶ Concluding Remarks

Strategic planning for project management, combined with good project processes, can compress time, cost, and quality initiatives. However, there are still critical decisions that must be made. Marketing must decide what products to offer and which markets to serve. The information systems people must assist in the design, development, and/or selection of support systems. And senior management must provide sufficient, qualified resources.

Strategic planning for excellence in project management needs to consider all aspects of the company: from the working relationships among employees and managers and between staff and management, to the roles of the various players (especially the role of executive project sponsors), to the company's corporate structure and culture. Other aspects of project management must also be planned. Strategic planning is vital for every company's health. Effective strategic planning can mean the difference between long-term success and failure. Even career planning for individual project managers ultimately plays a part in a company's excellence, or its mediocrity, in project management.

CHAPTER TWO

The Need to Plan for Project Management Maturity

▶ Introduction

In today's business world, project management has become the primary vehicle for achieving strategic objectives, realizing business benefits, and creating business value. Project management is treated in many firms as one of the four or five strategic competencies, necessary for the long-term survival of the firm, rather than as just another career-path position. Therefore, it is expected that companies will look for project management maturity models to help them become better at using project management delivery systems. Unfortunately, this is not as easy as it appears. More than 30 project management maturity models (PMMMs) are in the marketplace. There is some commonality among several of the models, but there are also significant differences. Selecting the right PMMM for your industry and type of business requires careful consideration.

Companies can grow in project management practices by capturing lessons learned and best practices. Most companies use internally generated lessons learned and best practices to improve on the processes, forms, guidelines, templates, and checklists that make up the project management methodology. PMMMs can achieve the same effect and identify windows of improvement opportunities at a faster rate. PMMMs can also be used to measure improvements in both tangible and intangible assets.

▶ The Need for a PMMM

The purpose of the PMMM is to assess the execution of the delivery system, seek out areas for improvement, establish a continuous improvement baseline, and then reassess performance periodically to see if continuous improvements were implemented. The results of the PMMM study could indicate changes that need to be made to project management processes as well as changes needed in the company's infrastructure. The results could indicate that more rather than less governance is needed. Since the strategic goals and objectives, as well as the delivery system, may be unique for each company, the PMMMs may need to have the capability for customization.

It is unrealistic to believe that PMMM assessments need to be conducted just once and that changes will take place in giant steps and occur quickly. Periodic reassessments of maturity allow you to determine that continuous improvements are taking place and whether gaps discovered in previous assessments have been closed. It is best that the changes take place in small steps rather than risking major disruptions to ongoing business or finding significant resistance when people are rapidly removed from their comfort zones.

Assessments and accompanied continuous improvement activities can provide the company with a competitive advantage. However, unless assessments are made periodically, it is unlikely that the competitive advantage will be sustained over the long term.

Assessments should be made with participation from all levels of management. Some models mistakenly focus entirely on executive management assessments. Lower and middle levels of management provide the staffing for projects and have issues that need to be considered. Although some people believe the project manager is solely responsible for the success or failure of a project, true success and failure should be shared between the project team as well as the functional managers who staffed the projects and committed to deliverables and meeting constraints. The results of the assessments can therefore be recommended changes that need to be made at all levels of management.

The purpose of conducting PMMM assessments is not only to improve the project delivery system but also to improve the deliverables and outcomes of the system. Assessments therefore mandate the use of performance metrics that can measure improvements in both tangible and intangible values. Most people believe that PMMM measurements focus exclusively on the standardization of the processes. However, tangible and intangible business value metrics must also be established that can be related to business performance or project execution (i.e., process) performance. Typical business performance maturity metrics might include:

- Customer satisfaction

- Customer acceptance

- Met quality guidelines

- Product performance level

- Launched on time

- Speed to market

- Met revenue goals

- Met unit sales goals

- Revenue growth

- Attain margin goals

- Attain profitability goals

- Internal rate of return (IRR) / return on investment (ROI)

- Development cost
- Breakeven time
- Met market share goals
- Percent of sales by new products

All of these bullets are related to strategic business objectives. Therefore, the PMMM must contain metrics that measure the ability of projects to be aligned either directly or indirectly with strategic business objectives.

Intangible value improvements may be better communications, better cooperation, more trust given to the teams, improved teamwork, and more involved corporate governance. Intangible values may be difficult to measure, but they are not immeasurable.

▶ Other Purposes for the PMMM

The PMMM serves other purposes than just seeking out project management maturity. Since programs are composed of several projects, the PMMM can also be used to look for continuous improvement efforts related to program management. Another option is portfolio management. A typical portfolio contains several projects. Project management maturity can convert a portfolio of projects into strategic business outcomes that meet strategic business objectives. Customization of the assessments may be necessary.

The PMMM described in this book is heavily oriented toward the behavioral factors that influence project management maturity. As such, the assessment questions can be customized to link organizational behavior in project management to traditional and nontraditional leadership models.

Leadership theories and models have been developed to study the actions and behaviors of successful leaders. In a project management environment, leadership does not rest with one person but rests on the behaviors of the project team and project governance groups. The focus is on how the project team accomplishes its goal rather than who has been formally assigned a leadership role.

John Adair[1] noted the following eight key functions for which team leaders are responsible. (Examples are given in brackets.) These key functions can be looked at with the assessment questions:

1. **Defining the task** (by setting clear objectives)
2. **Planning** (by looking at alternative ways to achieve the task and having contingency plans in case of problems)
3. **Briefing the team** (by creating the right team climate, fostering synergy, and making the most of each person's capabilities through knowing them well)
4. **Controlling what happens** (by being efficient in terms of getting maximum results from minimum resources)

---

[1] John Adair, *Action-Centered Leadership* (New York: McGraw-Hill, 1973), www.johnadair.co.uk/profiles.html.

5. **Evaluating results** (by assessing consequences and identifying how to improve performance)

6. **Motivating individuals** (by using both external motivators such as rewards and incentives as well as eliciting internal motivators on the part of each team player)

7. **Organizing people** (by organizing yourself and others using effective time-management practices, personal development, and delegation)

8. **Setting an example** (by recognizing that people observe their leaders and copy what they do).

► Defining Project Management Maturity

There is no universally accepted definition for *project management maturity*. This holds true whether the organization is project-driven or non-project-driven. As such, no single PMMM satisfies the needs of all companies.

In most companies, executive management establishes the strategic goals and objectives for the firm. These strategic goals and objectives help provide guidance in defining maturity in business terms such as competitiveness, growth in market share, and profitability. Project management is the delivery system for achieving the goals and objectives and is therefore treated as a subset of strategic planning activities. Unfortunately, many organizations do not recognize the importance of continuous improvements in project management and redefining maturity until they either have difficulty achieving their strategic goals and objectives or discover that performance reporting is not giving them accurate or timely information concerning completion dates and costs.

Every company has its own definition of *maturity*. Some definitions that are more closely aligned with business needs may include these:

- Compliance with project success criteria
- Completing work within the competing constraints
- Meeting strategic business goals and objectives
- Aligning project, program, and portfolio performance to strategic business objectives
- Effectively managing beneficial changes as part of continuous improvement efforts
- Maintaining or improving customer and stakeholder satisfaction
- Improving efficiency and effectiveness in execution
- Improving the organization's governance structure
- Improving how the firm competes in the marketplace

Project management maturity can be defined in generic terms:

> Project management maturity is the ongoing process of periodically identifying, measuring, implementing, and reassessing continuous improvement opportunities in the project delivery system and supporting infrastructure such that the organization can improve its ability to meet its strategic goals and objectives.

If we dissect the definition and look at some of the words closely, we find the following:

- *"ongoing process":* Project management maturity is not a single point in time. It is a continuous improvement process where there are always opportunities for doing things better. The purpose of the assessment is to identify these opportunities.

- *"reassessing":* It is necessary to perform reassessments periodically because, once again, maturity is an ongoing process. The time between reassessments is based on the cost of doing the assessments, along with the time needed between assessments to implement the continuous improvement opportunities.

- *"project delivery system and supporting infrastructure":* Although maturity can be sought out in all aspects of the way the firm conducts its business, the focus of the PMMM is to look mainly at the project delivery system and the necessary governance requirements that flow through the firm's infrastructure.

- *"meet its strategic goals and objectives":* If a company wants to stay in business, it must continuously meet or improve on its strategic goals and objectives. Periodic maturity assessment models can make this happen by identifying organizational strengths and weaknesses, areas for improvement, and ways to correct things that are not being done well; and benchmarking against industry competitors and world-class companies.

PMMMs are assessments against predefined or preselected targets. The targets might be the latest version of the *PMBOK® Guide*, industry standards, or industry or world-class benchmarking studies. Measurements can also be made against a company's definition of maturity.

Whatever the measurement target is, the expectations of meeting 100 percent of the target may be unrealistic. For example, some project managers may not need to know the types of contracts or contract management practices because those do not relate to their job now or in the future. Assessing them on this material may create a false impression of the maturity level of the organization.

▶ Advantages of Using a PMMM

Using PMMMs provides several advantages. They include the following:

- Identifies the organization's strengths and weaknesses
- Provides benchmarking information
- Identifies opportunities for continuous improvements
- Compares results against project management standards
- Compares results against industry standards
- Compares results against world-class standards

PMBOK is a registered mark of the Project Management Institute, Inc.

- Recognition that project management is a strategic rather than a tactical asset
- Allows a company to focus on competitive advantage opportunities rather than just improvements in time, cost, scope, and customer service
- Allows a company to improve its performance indicators
- Allows a company to identify metrics that will measure both tangible and intangible asset growth
- May raise awareness of how improvements in project management can benefit the organization

Some people argue that the intent behind using PMMMs is to encourage organizations to manage their projects better. While this is true, it also forces organizations to look at how they manage their programs and portfolios and provide governance and overall business practices. Therefore, there may be no one correct road map for how continuous improvement activities should take place, because the benefits apply to programs and portfolios as well as the project delivery system.

▶ Disadvantages of Using a PMMM

Not all PMMMs satisfy all the needs of a company. Some of the common disadvantages among the models may include these:

- Too much focus on process maturity
- Inability to focus on human resource issues such as management competencies, worker skill levels, capacity planning, and organizational development factors
- Too much focus on improvements in tangible assets, and little or no consideration of intangible assets
- Lack of flexibility, and inability to be customized to a given company's needs
- Inability to determine the impact that technology has on performance measurements and reporting as well as continuous improvement efforts
- A belief that the PMMM will clearly identify cause-and-effect relationships for continuous improvements
- A heavy focus on problem identification, and a weak focus on problem solving
- Assessments targeted for the wrong people in the company, especially those that cannot improve the project management delivery system
- A belief that the PMMM assessment needs to be performed only once
- The cost of using the PMMM, which may prevent companies from performing periodic reassessments
- Models that are highly complex
- Using a PMMM that is applicable only to project-driven organizations
- Lack of ability to benchmark against other companies and industries
- Levels of project management maturity that do not overlap

- Too many levels of maturity
- No support for program and portfolio management
- Too difficult to understand
- Too much focus on the author's definition of maturity
- Too much focus on a limited number of business objectives, such as ROI only

▶ Selecting a PMMM

With more than 30 PMMMs in the marketplace, it may be difficult for a company to select the right fit for the firm. The advantages and disadvantages can vary with the use of each PMMM. Some characteristics that are often used in differentiating between PMMMs include these:

- The number of maturity levels
- Whether it is based on a definition of maturity
- The standards against which it was designed
- The option to benchmark results against other users of this model
- The model's relationship to strategic planning activities
- Its applicability to projects, programs, and portfolio management practices
- Ability to identify the organization's strengths and weaknesses
- Ease with which the assessments can be made frequently
- Cost of performing the assessments
- Easy of understanding
- Ability to be customized

▶ Changing the Strategic Direction

In general, it isn't until the project management delivery system becomes critical because of deficient performance that organizations recognize the need for assessments. The results of an assessment may bring to light that the company is not doing things correctly and that changes are needed to improve competitiveness. As such, there must be an executive-level commitment to use the PMMM and take actions if needed.

The need for the assessment may be internally driven for improvements in efficiency and effectiveness, or customer driven. When customers are more mature than contractors, there is a risk that the customer will dictate to the contractor how the project should be managed. Also, as part of competitive bidding, customers may seek out the contractor that appears the most mature in project management and project delivery. When customers are pleased with a contractor's level of maturity in project management, they may treat the contractor as a strategic supplier and award the contractor with sole-source contracts.

▶ Maturity and Core Competencies

Reaching some level of maturity may appear on the surface to be a success. However, implementing any changes needed may enhance or destroy existing competencies that a firm uses to implement its day-to-day operations and execute its strategy. Simply stated, beware what you wish: some improvements can cause organizational disruption or even failure if employees are removed from their comfort zones and resist the changes.

Some changes resulting from assessments may be simple, such as changes to forms, guidelines, templates, and checklists. However, other changes may require that people be removed from their comfort zones and change management or transformational project management activities may be necessary.

There can be costs associated with seeking maturity and continuous improvement activities if organizational change management is needed. Typical costs may be:

- Hiring and training new recruits
- Changing the roles of existing personnel and providing training
- Relocating existing personnel
- Providing new or additional management support
- Updating computer systems
- Purchasing new software
- Creating new policies and procedures
- Renegotiating union contracts
- Developing new relationships with suppliers, distributers, partners, and joint ventures

▶ Maturity and Assessment Timing

The definition of maturity is usually made up of multiple measures, and each measure may have to be made at a different time along the entire PMMM initial measurement and reassessment period. With any maturity model based on levels of maturity, there is always the question of how long it will take to complete a given level of maturity. The answer is dependent on how long it takes to acquire the necessary knowledge for that level.

As shown in Figure 2.1, actual learning takes place in three areas: on-the-job experience, education, and knowledge transfer. On-the-job experience and knowledge transfer can take a great deal of time. Education is the fastest way to acquire the necessary knowledge, provided the right educational programs are selected.

Ideal project management knowledge would be obtained by allowing each employee to be educated on the results of the company's lessons-learned studies, including risk management, benchmarking, and continuous improvement efforts. This assumes that a project management office (PMO) is in place with responsibility for benchmarking, capturing lessons learned, and continuous improvement efforts (i.e., Levels 4 and 5 of the PMMM, as explained in Chapter 4). Unfortunately, for many companies, this is rarely achieved, and ideal learning is hardly ever reached.

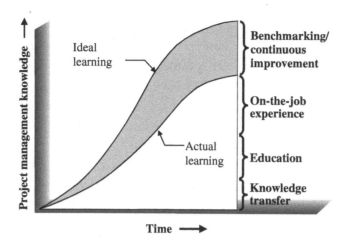

Figure 2.1 Project management learning curve.

Some companies establish maturity success criteria that indicate established milestones when certain levels of maturity are expected. Success criteria should be used because there is a significant difference between short-term and long-term success. As an example, a company develops a new product and quickly introduces it into the marketplace. Profitability in the short term may be difficult to assess because the company must first recover its product-development and market-introduction costs. In another example, a company incurs debt and constructs an attraction or a theme park. Short-term success may be the ability to service the debt, whereas long-term success may be paying off or reducing the debt load.

While customer acceptance may be measurable quickly, financial success criteria require long-term measurements. Many companies tend to focus on short-term profits at the expense of long-term growth. Most studies do not distinguish between short-term and long-term innovation project success.

▶ The Importance of Intangible Maturity Metrics

Companies can progress toward maturity by measuring both the tangible and intangible value that is being created. Metrics are measurements that are needed to validate that progress is being made. It is difficult, if not impossible, to manage what you do not measure. Mismeasurement is a root cause of mismanagement, especially when you measure just to collect data rather than measuring those items that make a difference. Simply stated, you must measure what matters. Without good metrics, it is difficult to determine if you are heading in the right direction, and you may be unsure as to what you can and cannot do to reach higher levels of maturity.

Perhaps the greatest change in project management as companies mature is the need for more complex metrics. For more than five decades, organizations have focused heavily on earned value measurement techniques stressing only the measurements of time, cost, and scope, because they were well understood and the easiest metrics to measure. Historically, when managing a project, things tended to move slowly, and companies adopted an attitude of "let's just wait and see what happens." Today, organizations must react more quickly than in the past, and these three metrics alone are insufficient.

Today, measurement techniques have advanced to the point that we believe we can measure anything. Projects now have both financial and nonfinancial metrics, and many of the nonfinancial metrics are regarded as intangible metrics. For decades, we shied away from intangibles. As stated by Nick Bontis,[2] "The intangible value embedded in companies has been considered by many, defined by some, understood by few, and formally valued by practically no one." Today, we can measure intangible factors as well as tangible factors that impact performance.

An *intangible asset* is nonmonetary and without physical substance. Intangible assets can be the drivers of innovation. In an excellent article, Ng et al. discuss various key intangible performance indicators (KIPIs) and how they can impact project performance measurements. The authors state,

> [T]here are many diverse intangible performance drivers which impact organizational [and maturity] success such as leadership, management capabilities, credibility, innovation management, technology and research and development, intellectual property rights, workforce innovation, employee satisfaction, employee involvement and relations, customer service satisfaction, customer loyalty and alliance, market opportunities and network, communication, reputation and trust, brand values, identity, image, and commitment, HR practices, training and education, employee talent and caliber, organizational learning, renewal capability, culture and values, health and safety, quality of working conditions, society benefits, social and environmental, intangible assets and intellectual capital, knowledge management, strategy and strategic planning and corporate governance.[3]

Today, there are measurement techniques for these. In dynamic organizations, both key performance indicator (KPIs) and KIPIs are used to validate performance.

There is resistance to measuring intangibles:

- Companies argue that intangibles do not impact the bottom line.
- Companies are fearful of what the results will show.
- Companies argue that they lack the capability to measure intangibles.

Intangible elements are now considered by some to be more important than tangible elements. This is happening with IT projects where executives are giving significantly more attention to intangible values. The value of intangibles can have a greater impact on long-term considerations than short-term factors. Management support for the value measurements of intangibles can also prevent short-term financial considerations from dominating decision-making. Intangibles are long-term measurements, and most companies seem to still focus on the short-term results.

---

[2] Nick Bontis, "Linking Human Capital Investment with Organizational Performance." *Drake Business Review* 1(2) (2002): 9.

[3] H. S. Ng, D. M. H. Kee, and M. Brannan, "The Role of Key Intangible Performance Indicators for Organisational Success." In *Proceedings of the 8th International Conference on Intellectual Capital, Knowledge, Management & Organizational Learning* (The Institute for Knowledge and Innovation Southeast Asia [IKI–SEA] of Bangkok University, 27–28 October 2011), vol. 2, pp. 779–787. Reading, UK: Academic Publishing Ltd., 2011.

CHAPTER THREE

Customizing the PMMM

▶ The Need for PMMM Customization

When companies make the decision to use a PMMM, three important assumptions are usually made:

1. Continuous improvement opportunities will be easy to identify.
2. The opportunities will be easy to implement.
3. Everyone will agree to and use the changes.

Although these assumptions seem reasonable, they are often difficult to achieve if the organization's project management approach is not aligned to the standards against which the model was created, or if the organization is primarily in a governance role for the coordination of several outsourced projects where each project is managed by a separate contractor that may have its own approach to managing projects. This situation holds true for many federal and state government agencies.

There are many situations in which customization would be beneficial. This should not imply that standard models have limited value. Standard models may be used if many of the assessment questions are aligned to the way the firm manages projects.

▶ Understanding Customization

PMMMs should assess the organization on the tools, techniques, processes, governance requirements, and desired human behavior that the organization deems realistic according to the organization's definition of maturity. The definition of maturity can and will change based on continuous improvements made and information obtained from benchmarking studies. Therefore, it may be best to use a PMMM that allows for both customization and updating of assessment instruments, even if initially the firm does not believe that customization is necessary.

The starting point in customization of the assessment instruments is senior management's vision about how they would like to run the company. The customization should also include the competencies that project personnel are expected to possess as part of the vision. These competencies are best established by a project management office (PMO) with guidance from any project management standards that exist. The PMO will have, in addition to possibly other assignments, responsibility for compliance with the firm's project management approach.

Although the light at the end of the tunnel is long-term efficiency and effectiveness in execution of activities, the customization should be made based on short-term expectations so that progress can be measured during reasonable assessment intervals. If the expectations are long term, and people must remove themselves from their comfort zones rapidly for change to occur, there could be significant resistance to the use of PMMMs. The interval between assessments should allow for deficiencies to be discovered and corrected using continuous improvement efforts and change-management practices, if necessary.

If the assessments are being made quarterly, then updating the assessment questions yearly may be the best approach. This will give the firm time to measure whether changes are taking place as well as the rate at which continuous improvements are being made. This also allows for obtaining organizational buy-in in small steps.

There are also valid reasons for using a standard assessment instrument without any customization. The primary benefit is that some companies that have developed a standard PMMM may allow you to benchmark your findings against other companies in their database that are in the same industry, are the same size, or have other similarities.

There are also disadvantages to each approach. Some people argue that with a customized PMMM, it may be impossible to benchmark against other companies. However, this is not always true. Even with a customized PMMM, you can benchmark against similar assessment results in a database. It is very difficult for a company to state how mature it is without some form of benchmarking, regardless of whether you benchmark all the results in the assessments or just part of them.

Simply stating that your company is mature may imply that there is no further room for improvement. Benchmarking at least part of a customized PMMM allows for identification of areas of improvement.

The cost of frequent customization becomes an issue. Therefore, companies that prefer a PMMM that allows for customization should select a model that is easy to use and has low assessment costs. Standardized assessments have the risk of including topics that are not applicable to your company and therefore may provide you with a false impression as to your firm's true maturity level.

▶ Issues with Public-Sector Project Management Maturity

It is not uncommon for both federal and state government agencies to outsource 90 percent or more of their projects for completion to private-sector firms but within the public-sector overview. To make matters worse, many of the contracts are awarded to the lowest bidder regardless of the bidder's project management expertise. Government agencies must maintain a project management system that can be used to manage their own internal projects and, at the same time, coordinate the project management efforts of a multitude of contractors that have their own forms of project management. In addition to having a great deal of ambiguity and uncertainty in the public sector, more so than in the private sector, political pressure and stakeholder management issues can create a negative effect on project management practices.

There are significant differences between project management in the public and private sectors. Although some of the issues can occur in both the public and private sectors, the impact on the public sector can be more severe. Some of the differences specifically related to the public sector when compared to the private sector include the following:

- There are significantly more stakeholders in the public sector.
- Political stakeholders have a great deal of instability and can change assignment because of elections, thus leading to changes in project objectives and financial expectations.
- Getting all the stakeholders to agree on goals and objectives may be difficult.
- Stakeholders may have limited knowledge about how project management should work.
- If a project is in trouble, government stakeholders may leave the project stranded and bail out to protect their political ambitions.
- It is a frequent practice to hide the root cause of a failure to protect the career path of the workers.
- Objectives are often established based on political agendas that are hidden.
- Political adversaries with their own hidden agendas may demand to be treated as active stakeholders.
- There is extensive coverage by the news media.
- Project managers may need support from agencies that are not part of the project team.
- There is fierce competition for limited resources due to political intervention, and recruitment practices are often more complex and time-consuming.
- During election years, politicians may arbitrarily reassign resources to their "pet projects" to increase their chances of reelection.
- There may be a lack of continuity and understanding of project management throughout the agency.
- Private-sector project managers may not be able to remove public-sector team members who are underperforming.
- Public-sector team members may see the assignment to the project as a secondary job that may not have an impact on their performance reviews.
- There may not be formal job descriptions or career paths established for project managers.

- Regardless of how the private-sector project manager desires to manage the project, public-sector team members may have to follow inflexible government policies and rules that often create more work.

- The success of public-sector projects may not be measured until well into the future.

- The definition of success is significantly more complex, and there may be an abundance of constraints.

- Most projects in the public sector do not have well-defined success or failure criteria.

- Government agencies generally function in a hierarchical environment and without any type of cooperative culture.

- New risks and uncertainties occur more quickly than in the private sector.

- Risk management practices may be avoided for fear of exposing risks that could lead Congress to cancel the project, in which case project personnel would have blemishes on their record.

- There may be a lack of buy-in for the business case, and people in senior positions may be unwilling to pull the plug for fear it will impact their career.

- Since politicians in the public sector do not like to hear unwelcome news, the number of metrics used to measure project performance is minimized.

- Although most budgets are tighter than in the private sector, the funding for cost overruns is paid for by future generations.

- Public-sector project managers and team members must follow the chain of command for information they need and for decision-making support.

- Public sector project managers have less flexibility for tradeoffs and are forced to work within the existing constraints and goals.

- Cross-boundary communication and cooperation can be difficult, thus making the implementation of matrix management difficult.

- The public sector tends to move slowly when the outcome of a project mandates change management.

There are many reasons public-sector projects fail, and quite a few are the result of the previous bullets. They include:

- Inability to identify the key stakeholders
- Optimistic schedules with no contingency plans for late deliverables
- Insufficient or unqualified resources
- Not enough time devoted to up-front project planning
- Constantly changing priorities due to political agendas
- Inability to get an agreed-on prioritization of activities
- Poor risk management practices for fear of exposing the seriousness of risks
- No revalidation of assumptions and constraints
- Lack of repeatable project management processes

- Inexperienced project managers
- Failing to benefit from, or capture, lessons learned and best practices from projects

All government agencies, even the smallest municipalities, need some form of project management. Progress is being made regardless of the obstacles. Some government agencies have established life cycles for externally as well as internally managed government projects, but there is still a heavy reliance on contractors for project management expertise. In some government agencies, project management is now a career-path position rather than a part-time assignment. There is still difficulty in some organizations defining the skills needed to be a project manager in the government so that it could become a career-path position. Skills needed include some capabilities in management, human resources, organization, and dealing with political factors.

At the end of each project, government agencies have traditionally focused heavily on the deliverables of the project with very little interest in identifying the best practices and lessons learned that were discovered by the contractors using government-funded projects. Today, some government agencies have established PMOs to capture these best practices, and some of the government's requests for proposals state that all best practices and lessons learned must be shared with the sponsoring government agency at project completion. The use of PMOs in the government in becoming common. The PMOs are responsible for improving project management capabilities and monitoring compliance.

As government agencies try to manage more projects internally, there is an increasing need to build up public trust in the government. Effective project management practices can make this happen. Therefore, use of PMMMs in government agencies should take place regularly. But given the differences mentioned previously between the public and private sectors, customization of PMMM assessment instruments should be considered to address the differences and causes of failure. Traditional PMMMs can be used, and benchmarking is an option, but there is still uncertainly as to whether issues related to maturity can be assessed without some customization.

▶ Olympic Games Project Management Maturity

For a little more than two weeks every other year, we watch the heroics of some of the greatest athletes in the world as they compete at the Summer and Winter Olympic Games. Over 13,000 athletes from more than 200 countries compete in more than 30 different sports and nearly 400 events. Many of the athletes have prepared for years for an event that is measured in seconds or minutes. In some cases, the Olympics and its media exposure provide unknown athletes with the chance to attain national and sometimes international fame.

But while the Olympic Games appear to last slightly over two weeks, many people do not realize the complexities of preparing for such events or the activities that follow after the Games have ended. This is where project management plays a key role. The preparation time for the Games can be more than a decade prior to the opening ceremonies, and business-related activities that are part of the Olympic Games can go on for years after the Games have ended. There can be hundreds of contractors involved in preparation for the Games, and each one may have its own approach to project management.

Most of the projects, based on which city hosts the Games, are likely infrastructure projects organized by the host city's government in coordination with the National and International Olympic Committees. Typical infrastructure projects are:

- Sports facilities
- Housing for athletes, officials, and tourists
- Communications systems and IT
- Facilities for the media and the press
- Railways, stations, bridges, tunnels, interchanges, roads, and airports
- Power stations
- Traffic and crowd management
- Security

The city hosting the Games may have a standard project management approach that it has developed and that is used by local contractors. The approach may be for all projects including infrastructure projects, not necessarily those just for the Olympic Games. Many local contractors may accept the host city's approach and the accompanying continuous improvement efforts since they expect other contracts over several years. But the Olympic Games may occur just once in a lifetime in this city, and it is therefore unlikely that one-time contractors will modify their approach to appease the host.

▶ Capturing Olympic Games Lessons Learned

The Olympic Games tend to cost the host city billions of dollars more than anticipated. As the funding required to support the Games grows, so does the need to capture best practices and lessons learned from previous Olympic Games. Repeating mistakes can be costly.

Perhaps the most important lesson learned from the Montreal Olympics in 1976, Athens in 2004, Salt Lake City in 2010, and London in 2012 was the need for a collaborative working environment. Open, meaningful communication is essential. This requires a willingness to hear both good and unwelcome news.

When politicians create a culture in which they refuse to hear unwelcome news, the result is a breakdown in communications between politicians and various Olympic stakeholder groups. This often leads to a slowdown on projects because problem resolutions are beyond the project manager's level of authority. Action items can then remain in the system forever without receiving proper attention.

Adding to that, an important lesson learned from Athens was the necessity for open and honest communication among stakeholders, including government stakeholders. For the 2010 Olympics in Salt Lake City, governance offices were created, including the use of project charters. Authority and decision-making responsibilities were decentralized to various teams in order for rapid decision-making to take place to keep the program moving. Information transmitted between the public and private sectors must be in a language that is easy for everyone to understand so they can therefore make informed decisions based on evidence rather than guesses. This accelerates the handling of action items.

Several other project management–related lessons were learned from past Olympics. They include the following:

- There must be transparency about all assumptions made and how the costs in the bid were derived.
- Critical issues, including unwelcome news, should be brought to the surface for resolution rather than being buried.
- There are limited opportunities for tradeoffs other than an escalation of costs.
- Given that economic conditions can change over the life cycle of the Olympics, continuous revalidation of assumptions is necessary.
- Because of the length of the life cycle, risk-management activities should focus on mitigating future damages.
- The sharing of information is critical.
- There must be a clear understanding of each party's role, responsibility, and decision-making authority. This include governance personnel.
- Stakeholders and governance personnel must understand project management and how their actions and decisions can impact the performance of the projects.
- Project managers must know how to deal with fraudulent activities, collusion, kickbacks, embezzlement, and influence peddling. The project manager may not be able to control these without support from above.

Best practices and lessons learned are now being captured for Olympic events. Rather than call it a best practices library (as is common in industry), it is referred to as the *Learning Legacy*. After the London Olympics, seminars were taught about lessons learned. These seminars are expected to continue after each Olympics. Hopefully cities that plan to host the Games in the future will include these lessons learned in their project management delivery systems. These best practices and lessons learned can be used to customize a host city's assessments of project management maturity, perhaps during the bidding stage. This could then allow ample time for changes to be made and possibly minimize damage from cost overruns.

CHAPTER FOUR

An Introduction to the Project Management Maturity Model (PMMM)

▶ Introduction

All companies desire to achieve maturity and excellence in project management. Unfortunately, not all companies recognize that the time frame can be shortened by performing strategic planning for project management. The simple use of project management, even for an extended amount of time, does not necessarily lead to excellence. Instead, it can result in repetitive mistakes and, what's worse, learning from your own mistakes rather than from the mistakes of others.

Strategic planning for project management is unlike other forms of strategic planning in that it is most often performed at the middle-management level, rather than by executive management. Executive-level management is still involved, mostly in a supporting role, and provides strategic information and funding together with employee release time for the effort. Executive involvement will be necessary to make sure that whatever is recommended by middle management will not result in unwanted changes to the corporate culture.

Today, because project management is seen as a strategic competency, executives are involved in project management maturity. Executives are expected to be champions and active participants in the project management process, and to drive the development and implementation process from the top down. In one automotive supplier, middle management developed an outstanding project management methodology. Senior management sponsored the implementation process to make sure that the entire organization bought in to the methodology and used it. After implementation, executive sponsorship diminished. This resulted in a very weak continuous improvement process because nobody was driving the change process from the top down. An executive champion was then reinstated, and continuous improvement flourished to the point where this supplier now has one of the best project management methodologies within the automotive industry.

Organizations tend to perform strategic planning for new products and services by laying out a well-thought-out plan and then executing the plan with the precision of a surgeon. Unfortunately, strategic planning for project management, if performed at all, is done on a trial-by-fire basis. However, there are models that can be used to assist corporations in performing strategic planning for project management and achieving maturity and excellence in a reasonable amount of time.

▶ The Foundation for Excellence

The foundation for achieving excellence in project management can best be described as the project management maturity model (PMMM), which comprises five levels, as shown in Figure 4.1. Each of the five levels represents a different degree of maturity in project management. Each level is discussed in detail in the remaining chapters. The levels are as follows:

- *Level 1—Common language:* In this level, the organization recognizes the importance of project management and the need for a good understanding of the basic knowledge on project management and the accompanying language/terminology. Not all companies agree on project management terminology. The terminology used in *A Guide to the Project Management Body of Knowledge (PMBOK® Guide)* is not the only acceptable terminology. Many companies that are quite successful in project management have their own terminology. Flexible project management approaches such as with frameworks as used in agile and Scrum have their own terminology.

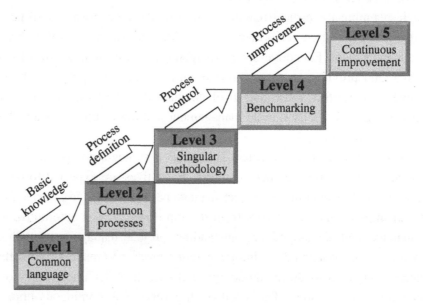

Figure 4.1 The five levels of project management maturity.

PMBOK is a registered mark of the Project Management Institute, Inc.

- *Level 2—Common processes:* In this level, the organization recognizes that common processes need to be defined and developed such that successes on one project can be repeated on other projects. Also included in this level is the recognition of the application and support of the project management principles to other methodologies employed by the company.

- *Level 3—Singular methodology:* In this level, the organization recognizes the synergistic effect of combining all corporate methodologies into a singular methodology, the center of which is project management. The synergistic effects also make process control easier with a single methodology than with multiple methodologies. Companies that reach Level 3 for traditional project management maturity may then desire to go to a flexible project management approach whereby each project manager can customize the tools for a given client.

- *Level 4—Benchmarking:* This level contains the recognition that process improvement is necessary to maintain a competitive advantage. Benchmarking must be performed on a continuous basis. The company must decide whom to benchmark and what to benchmark.

- *Level 5—Continuous improvement:* In this level, the organization evaluates the information obtained through benchmarking and must then decide whether this information will enhance the use of project management processes.

When we talk about levels of maturity (and even life-cycle phases), there exists a common misbelief that all work must be accomplished sequentially (i.e., in series). This is not necessarily true. Certain levels can and do overlap. The magnitude of the overlap is based on the amount of risk the organization is willing to tolerate. For example, a company can begin the development of project management checklists to support the methodology while it is still providing project management training for the workforce. A company can create a center of excellence (COE) in project management or a project management office (PMO) before benchmarking is undertaken.

▶ Overlap of Levels

Although overlapping does occur, the order in which the phases are completed cannot change. For example, even though Level 1 and Level 2 can overlap, Level 1 *must* still be completed before Level 2 can be completed. Overlapping of several of the levels can take place, as shown in Figure 4.2:

- *Overlap of Level 1 and Level 2:* This overlap will occur because the organization can begin the development of project management processes either while refinements are being made to the common language or during training.

- *Overlap of Level 3 and Level 4:* This overlap occurs because, while the organization is developing a singular methodology, plans are being made as to the process for improving the methodology.

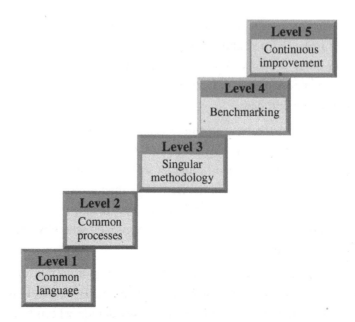

Figure 4.2 Overlapping levels.

- *Overlap of Level 4 and Level 5:* As the organization becomes more and more committed to benchmarking and continuous improvement, the speed at which the organization wants changes to be made can cause these two levels to have significant overlap. The feedback from Level 5 back to Level 4 and Level 3, as shown in Figure 4.3, implies that these three levels form a continuous improvement cycle, and it may even be possible for all three of these levels to overlap.

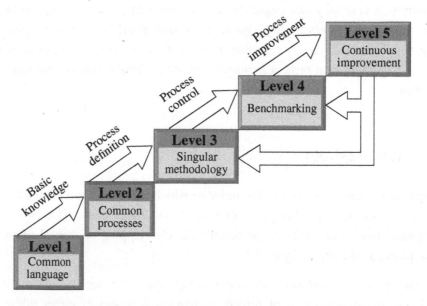

Figure 4.3 Feedback among the five levels of project management maturity.

Level 2 and Level 3 generally do not overlap. It may be possible to begin some of the Level 3 work before Level 2 is completed, but this is highly unlikely. Once a company is committed to a singular methodology, work on other methodologies generally terminates.

Also, if a company is truly astute in project management, it may be possible to begin benchmarking efforts even as early as Level 1. This way the company may learn from the mistakes of others rather than from its own mistakes. It is possible for Level 4 to overlap all of the first three levels.

For example, a company recognized that project management would certainly be beneficial and began performing strategic planning for project management. Senior management approved the creation of a project management office (i.e., a component of Level 4 of the PMMM) to head up the strategic planning for project management effort. The PMO began functioning while the company began activities in Level 1 of the PMMM. In less than six months, the company was able to complete the first three levels of the PMMM.

▶ Risks

Risks can be assigned to each level of the PMMM. For simplicity's sake, the risks can be labeled as low, medium, and high. The level of risk is most frequently associated with the impact of having to change the corporate culture. Assigning risk is a subjective assessment of the way the corporate culture might react at a specific level of the PMMM. The following definitions can be assigned to these three risks:

- *Low risk:* There will be virtually no impact on the corporate culture, or the corporate culture is dynamic and readily accepts change.

- *Medium risk:* The organization recognizes that change is necessary but may be unaware of the impact of the change. Instituting multiple-boss reporting would be an example of a change carrying medium risk.

- *High risk:* High risks occur when the organization recognizes that the changes resulting from the implementation of project management will cause a change in the corporate culture. Examples include the creation of project management methodologies, policies, and procedures, as well as decentralization of authority and decision-making.

Level 3 has the highest risks and degree of difficulty for the organization. This is shown in Figure 4.4. Once an organization is committed to Level 3, the time and effort needed to achieve the higher levels of maturity have a low degree of difficulty. Achieving Level 3, however, may require a major shift in the corporate culture.

Everyone in the organization, perhaps even on a global basis, must use the same methodology for project management. This could easily lead to changes in work habits, social groups, and comfort zones. Unfortunately, developing a singular methodology, often referred to as an *enterprise project management methodology*, is not easy. Most people argue that several methodologies may be needed, such as one for information systems projects and another for new product development. Some companies that claim they are using a singular enterprise project management methodology may have the information systems methodology as a subset of the new product development methodology.

Level	Description	Degree of difficulty
1	Common language	Medium
2	Common processes	Medium
3	Singular methodology	High
4	Benchmarking	Low
5	Continuous improvement	Low

Figure 4.4 Degree of difficulty associated with each level of the PMMM.

Implementing an enterprise methodology multinationally creates additional problems. Each country can have its own laws, types of contracts, labor and employment requirements, and standards. Language barriers are also an issue.

The following chapters have detailed descriptions of each of the five levels of the PMMM. For each of the five levels of maturity, we discuss:

- The characteristics of the level
- What roadblocks exist that prevent companies from reaching the next level
- What must be done to reach the next level
- Potential risks

▶ Assessment Instruments

Also included in each of the next five chapters is an assessment instrument to help you determine your organization's degree of maturity at each level. No two companies implement project management the same way. Since maturity will be different from company to company, the questions in these assessments can be modified to satisfy the needs of individual companies. Simply stated, using the principles contained in each chapter, you can customize the assessment instruments for each level. Not all companies follow the *PMBOK® Guide*. And those that do follow it, do not necessarily emphasize all areas of the *PMBOK® Guide*.

The assessment instruments can be custom-designed to include company-specific areas of maturity such as supply chain management, the PMO, and portfolio management implementation. The assessment grading system can also be custom-designed to identify ways to speed up the implementation process. For additional information on customization, contact Lori Milhaven at the International Institute for Learning, 212-515-2121.

CHAPTER FIVE

Level 1: Common Language

► Introduction

Level 1 is the level in which the organization first recognizes the importance of project management. The organization may have a cursory knowledge of project management or simply no knowledge. There are certain characteristics of Level 1, as shown in Figure 5.1:

- If the organization is using project management at all, the use is sporadic. Both senior management and middle-level management may be providing meaningless or lip-service support to the use of project management. Executive-level support is nonexistent.

- There may exist small pockets of interest in project management, with most of the interest existing in the project-driven areas of the firm.

- No attempt is made to recognize the benefits of project management. Managers are worried more about their own empires, power, and authority, and appear threatened by any new approach to management.

- Decision-making is based on what is in the best interest of the decision-maker, rather than the firm as a whole.

- There exists no investment or support for project management training and education, for fear that this new knowledge may alter the status quo.

In Level 1, project management is recognized, as in all companies, but not fully supported. There is resistance to change, and some companies never get beyond this level.

The starting point to overcome the characteristics of Level 1 is a sound, basic knowledge of the principles of project management. Education is the name of the game to complete Level 1. Educational programs on project management cover the principles of project management, advantages (and disadvantages) of project management methodologies, and the basic language of project management.

Project management certification training courses are ideal to fulfill organizational needs to reach Level 1 of the project management maturity model (PMMM). Project management and total quality management (TQM) are alike in that both require

> ### Common language
> - Lip service to project management
> - Virtually no executive-level support
> - Small "pockets" of interest
> - No attempt to recognize the benefits of project management
> - Self-interest comes before company's best interest
> - No investment in project management training and education

Figure 5.1 Characteristics of Level 1.

an all-employee training program that begins at the senior levels of management. However, the magnitude of the training program and the material covered can vary, based on the type of employees, the skills needed, and the size and nature of the projects within the organization. Executives may require only an overview course of three to six hours, whereas employees who are more actively involved in the day-to-day activities of projects may require week-long training programs.

▶ Roadblocks

Training programs alone cannot overcome the fears and apprehensions that exist in the management ranks concerning the implementation of project management. Figure 5.2 illustrates the most common roadblocks that prevent an organization from completing Level 1.

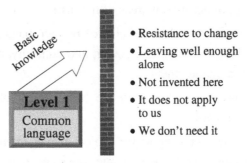

Figure 5.2 Roadblocks to completion of Level 1.

Resistance to change is the result of management's belief that the implementation of project management will cause "culture shock," where functional managers will have to surrender some or all of their authority to the project managers. As a result, numerous excuses will appear as to why project management is not needed or will not work. Typical comments include: "We don't need it." "It doesn't apply to our business." "Let's leave well enough alone."

The implementation of project management does not have to be accompanied by shifts in the power and authority spectrum. However, there may be a shift in the reporting structure, inasmuch as project management is almost always accompanied by multiple-boss reporting. All training programs on project management emphasize multiple-boss reporting.

▶ Advancement Criteria

Five key actions are required before the organization can advance to Level 2:

- Arrange for initial training and education in project management.
- Encourage the training (or hiring) of certified project management professionals (PMP® credential holders).
- Encourage employees to begin communicating in common project management language.
- Recognize available project management tools.
- Develop an understanding of the principles of project management: the project management body of knowledge as spelled out in the *PMBOK® Guide*.

The last item may prove the most difficult in non–project-driven organizations where project management is not regarded as a profession.

The successful completion of Level 1 usually occurs with a medium degree of difficulty. The time period to complete Level 1 could be measured in months or years, based on such factors as:

- Type of company (project-driven versus non–project-driven)
- Size and nature of the projects
- Amount of executive support
- Visibility of executive support
- Strength of the existing corporate culture
- Previous experience, if any, with project management
- Corporate profitability
- Economic conditions (inflation, recession, etc.)
- The speed by which training can be accomplished

▶ Risk

Level 1 carries a medium degree of risk. The organization might very well be resistant to change. Management may be fearful of a shift in the balance of power and authority.

Another major problem at Level 1 is when the organization first recognizes the complexities of multiple-boss reporting, which is a necessity for project management. Multiple-boss reporting can affect the wage and salary administration program and how employees are evaluated.

PMP and PMBOK are registered marks of the Project Management Institute, Inc.

Typical factors that cause Level 1 to present a medium level of risk include:

- Fear of organizational restructuring
- Fear of changes in roles and responsibilities
- Fear of changes in priorities

▶ Assessment Instrument for Level 1

Completion of Level 1 is based on gaining knowledge of the fundamental principles of project management and its associated terminology. The requirements for completing Level 1 can be fulfilled through a good understanding of the *PMBOK® Guide* prepared by the Project Management Institute (PMI)®.

Testing on the *PMBOK® Guide* is a good indicator of where you stand in relation to Level 1. The testing can be accomplished on an individual basis or by taking the average score from a group of individuals.

Following are 80 questions covering the *PMBOK® Guide* and the basic principles of project management. There are five answers for each question. Although some of the answers may appear quite similar, you must select one and only one answer. After you finish question 80, you will be provided with written instructions on how to grade the exercise.

▶ Questions

1. A comprehensive definition of scope management would be:
   A. Managing a project in terms of its objectives through all life-cycle phases and processes
   B. Approval of the scope baseline
   C. Approval of the detailed project charter
   D. The processes required to ensure that the project includes all the work required to complete the project successfully

2. The most common types of schedules include all but one of the following:
   A. Project network diagrams with date information added
   B. Resource-leveling heuristics
   C. Bar charts
   D. Milestones

3. The communications environment involves both internal factors and external factors. An example of a typical internal factor is:
   A. Power games
   B. Business environment
   C. Technical state of the art
   D. Political environment

4. The most effective means of determining the cost of a project is to price out the:

   A. Work breakdown structure (WBS)

   B. Linear responsibility chart

   C. Project charter

   D. Scope statement

5. Employee unions would most likely satisfy which level in Maslow's hierarchy of needs?

   A. Social

   B. Self-actualization

   C. Esteem

   D. Physiological

6. A document that describes the procurement item in sufficient detail to allow prospective sellers to determine if they can provide it is a:

   A. Contractual provision

   B. Statement of work (SOW)

   C. Terms and conditions statement

   D. Proposal

7. Future events or outcomes that are favorable are called:

   A. Risks

   B. Opportunities

   C. Surprises

   D. Contingencies

8. An example of an appraisal cost in terms of the cost of quality is:

   A. Surveys of vendors, suppliers, and subcontractors

   B. Evaluations of customer complaints

   C. Internal-external design reviews

   D. Process studies

9. Perhaps the biggest problem facing the project manager during integration activities within a matrix structure is:

   A. Coping with employees who report to multiple bosses

   B. Too much sponsorship involvement

   C. Unclear functional understanding of the technical requirements

   D. Escalating project costs

10. If you wish to compare actual project results to planned or expected results, you should:

   A. Hold a performance review

   B. Request a progress report

   C. Perform a trend analysis

   D. Perform a variance analysis

11. Communications has many different dimensions. Deciding to form a group among project managers in your organization to discuss lessons learned and best practices to follow is an example of which of the following dimensions:

    A. Internal

    B. External

    C. Horizontal

    D. Vertical

12. Which of the following methods is best suited to identifying the "vital few"?

    A. Pareto analysis

    B. Cause-and-effect analysis

    C. Trend analysis

    D. Process control charts

13. A collection of formal procedures that includes the steps by which official project documents will be changed is defined through:

    A. The project management information system

    B. The change control system

    C. The Change Control Board

    D. Performance reports

14. A risk is noted by having a cause and:

    A. If it occurs, it only has a negative effect on the project's objectives

    B. A known unknown

    C. If it occurs, it has a consequence

    D. A constraint

15. In general, differences between and among project stakeholders should be resolved in favor of the:

    A. Project sponsor

    B. Performing organization

    C. Functional manager

    D. Customer

16. Project life cycles share many common characteristics, which include all of the following except:

    A. Increased ability for stakeholders to influence the final characteristics of the project toward the end of the life cycle

    B. Probability of successful completion being lowest at the beginning of the project

    C. Reduced ability of stakeholders to influence the final cost of the project as the project continues

    D. A low staffing level at the start of the project

17. Smoothing out resource requirements from period to period is called:

    A. Resource allocation

    B. Resource partitioning

    C. Resource leveling

    D. Resource quantification

18. The difference between the EV (earned value) and the PV (planned value) is referred to as:

    A. The schedule variance

    B. The cost variance

    C. The estimate of completion

    D. The actual cost of the work performed

19. Project managers must use a number of different interpersonal influences on projects to contribute to project success. If the project manager is viewed as being empowered to issue orders, he or she is using which of the following types of power?

    A. Expert

    B. Reward

    C. Referent

    D. Legitimate

20. The sender-receiver model in project communications includes:

    A. The choice of media

    B. The feedback loops and barriers to communications

    C. The presentation and meeting management techniques

    D. The choice of technology

21. A deliverable-oriented grouping of project components to organize and define the total project scope is:

    A. A detailed plan

    B. A linear responsibility chart

    C. A work breakdown structure (WBS)

    D. A cost accounting coding system

22. Modern quality management and project management are complementary because both disciplines recognize the importance of all but one of the following:

    A. Customer satisfaction

    B. Processes within phases

    C. Management responsibility

    D. Inspection over prevention

23. In which of the following circumstance(s) would you be most likely to buy goods or services instead of producing them in-house?

   A. Your company has excess capacity and can produce the goods or services.

   B. Your company lacks capacity.

   C. There are many reliable vendors for the goods or services that you are attempting to procure, but the vendors cannot achieve your level of quality.

   D. Your company has an ongoing need for the item.

24. A limitation of the bar chart is:

   A. Difficulty changing it once it is prepared

   B. Difficulty in understanding if you do not have a knowledge of project management

   C. Difficulty adding new items to it as the project changes

   D. Difficulty performing any sensitivity analysis because it does not show the uncertainty involved in performing activities

25. The tool and technique used for risk-management planning is:

   A. Assessment of stakeholder risk tolerances

   B. Planning meetings

   C. Interpersonal and team skills

   D. Assumption and constraint analyses

26. Typically, during which phase in a project life cycle are most of the project expenses incurred?

   A. Concept phase

   B. Development or design phase

   C. Execution phase

   D. Termination phase

27. Going from Level 3 to Level 4 in the work breakdown structure (WBS) will result in:

   A. Less estimating accuracy

   B. Better control of the project

   C. Lower status reporting costs

   D. A greater likelihood that some key project element has been overlooked

28. Conflict management requires problem-solving. Which of the following is often referred to as a problem-solving technique and used extensively by project managers in conflict resolution:

   A. Confrontation

   B. Compromise

   C. Smoothing

   D. Forcing

29. Estimating the effect of the change of one project variable on the overall project is known as:

    A. The project manager's risk-aversion quotient

    B. The total project risk

    C. The expected value of the project

    D. Sensitivity analysis

30. Power games, withholding information, and hidden agendas are examples of:

    A. Feedback

    B. Communication barriers

    C. Indirect communication

    D. Mixed messages

31. The basic terminology for networks includes:

    A. Activities, events, personnel, skill levels, and slack

    B. Activities, documentation, events, personnel, and skill levels

    C. Slack, activities, events, and time estimates

    D. Time estimates, slack, sponsorship involvement, and activities

32. The "control points" in the work breakdown structure (WBS) used for assignments to specific organizational units or individuals are:

    A. Work packages

    B. Subtasks

    C. Tasks

    D. Code of accounts

33. Establishing a market window on a technology project or achieving government-mandated compliance with environmental remediation are examples of:

    A. Imposed dates

    B. Weather restrictions on outdoor activities

    C. Major milestones

    D. Product characteristics

34. An example of a constraint to consider during procurement planning is:

    A. Indirect costs

    B. Legal obligations and penalties

    C. Market conditions

    D. Procurement resources

35. The basic elements of a communication model include:

    A. Written and oral, and listening and speaking

    B. Communicator, encoding, message, medium, decoding, receiver, and feedback

    C. Reports and briefings as well as memos and ad hoc conversations

    D. Reading, writing, participating in meetings, and listening

36. Assume that you are managing a project that is a joint venture between your company and two other firms. The project's quality policy then should be:

    A. Your responsibility to prepare

    B. The same as that of your customer

    C. The same as that of your company

    D. Prepared by the project team

37. The three most common types of project cost estimates are:

    A. Order of magnitude, parametric, and budget

    B. Parametric, definitive, and top down

    C. Order of magnitude, definitive, and bottom up

    D. Order of magnitude, budget, and definitive

38. Good project objectives must be:

    A. General rather than specific

    B. Established without considering resource constraints

    C. Realistic and attainable

    D. Measurable, intangible, and verifiable

39. The process of determining which risks might affect the project and documenting their characteristics is:

    A. Risk identification

    B. Risk response planning

    C. Risk management planning

    D. Qualitative risk analysis

40. In which type of contract arrangement is the *contractor* most likely to control costs?

    A. Cost-plus-fixed fee

    B. Firm-fixed price

    C. Time and materials

    D. Fixed-price-incentive firm target

41. A project can best be defined as:

    A. A series of nonrelated activities designed to accomplish single or multiple objectives

    B. A coordinated effort of related activities designed to accomplish a goal without a well-established end point

    C. Cradle-to-grave activities that must be accomplished in less than one year and consume human and nonhuman resources

    D. Any undertaking with a definable time frame and well-defined objectives that consumes both human and nonhuman resources with certain constraints

42. Risk management decision-making falls into three broad categories:

    A. Certainty, risk, and uncertainty

    B. Probability, risk, and uncertainty

    C. Probability, risk event, and uncertainty

    D. Hazard, risk event, and uncertainty

43. A process is considered to be out of control when there are how many consecutive data points (minimum) on either side of the mean on a control chart?

    A. 3

    B. 7

    C. 9

    D. 11

44. The work breakdown structure (WBS), the work packages, and the company's accounting system are tied together through:

    A. The code of accounts

    B. The overhead rates

    C. The budgeting system

    D. The capital budgeting process

45. A program can best be described as:

    A. A grouping of related activities that lasts two years or more

    B. A collection of projects and other work designed to meet strategic business objectives

    C. A group of projects managed in a coordinated way to obtain benefits not available from managing them individually

    D. A product line

46. Which of the following types of power comes through the organizational hierarchy:

    A. Coercive, legitimate, referent

    B. Reward, coercive, expert

    C. Referent, expert, legitimate

    D. Legitimate, coercive, reward

47. The most common definition of project success is:

    A. Within time

    B. Within time and cost

    C. Within time, cost, and technical performance requirements

    D. Within time, cost, performance, and acceptance by the customer/user

48. Activities with zero time duration are referred to as:

   A. Critical path activities

   B. Noncritical path activities

   C. Slack time activities

   D. Dummies

49. The procurement planning process should be accomplished during:

   A. Scope definition

   B. Solicitation planning

   C. Project initiation

   D. Scope planning

50. Project cash reserves are often used for adjustments in escalation factors, which may be beyond the control of the project manager. Other than possible financing (interest) cost and taxes, the three most common escalation factors involve changes in:

   A. Overhead rates, labor rates, and material costs

   B. Rework, cost-of-living adjustments, and overtime

   C. Material costs, shipping costs, and scope changes

   D. Labor rates, material costs, and cost reporting

51. The critical path in a network is the path that:

   A. Has the greatest degree of risk

   B. Is the longest during the project and determines its duration

   C. Must be completed before all other paths

   D. Has activities with float greater than zero

52. The major difference between project and line management is that the project manager may not have any control over which basic management function?

   A. Decision-making

   B. Staffing the project

   C. Tracking/monitoring

   D. Reviewing project performance

53. During which phase of a project is the uncertainty the greatest?

   A. Design

   B. Development/execution

   C. Concept

   D. Closing

54. Quality often is confused with grade. This means that:

   A. Low quality is always a problem, but low grade may not be a problem.

   B. Low grade is always a problem, along with low quality.

C. Quality is defined as a category or rank, with entities having the same functional use but different technical characteristics.

D. Grade is defined as the total characteristics of an entity that bear on its ability to satisfy stated or implied needs.

55. Project managers need exceptionally good communication and negotiation skills primarily because:

A. They may be leading a team over which they have no direct control.

B. This need is mandated by the project's procurement activities.

C. They are expected to be technical experts.

D. They must provide executive/customer/sponsor briefings.

56. For effective communication, the message should be oriented to:

A. The initiator

B. The receiver

C. The management style

D. The corporate culture

57. Common factors that may constrain how the project team is organized include all but one of the following:

A. The structure of the performing organization

B. Preferences of the team

C. Expected staff assignments

D. Responsibility Assignment Matrix

58. On a precedence diagram, the arrow between two boxes is called:

A. An activity

B. A constraint

C. An event

D. The critical path

59. In which type of contract arrangement is the *contractor* least likely to control costs?

A. Cost-plus-incentive fee

B. Firm-fixed price

C. Fixed-price-award fee

D. Purchase order

60. The financial closeout of a project dictates that:

A. All project funds have been spent.

B. No charge numbers have been overrun.

C. No follow-on work from this client is possible.

D. No further charges can be made against the project.

61. A graphical display of accumulated costs and labor hours for both budgeted and actual costs, plotted against time, is called:

A. A trend line

B. A trend analysis

C. An S curve

D. A percent completion report

62. If you are using a control chart and find that the process is in control, it is important to recognize that:

A. The process should not be adjusted.

B. The process should not be changed to provide improvements.

C. Sources of random variation can be easily changed without the need to restructure the process.

D. Sources of random variation are never present.

63. The major difference between PERT and CPM networks is:

A. PERT requires three time estimates whereas CPM uses one time estimate.

B. PERT is used only for construction projects whereas CPM is used solely for R&D.

C. PERT addresses only time whereas CPM also includes costs and resource availability.

D. PERT is measured in days whereas CPM uses weeks or months.

64. Information can be shared by team members and other stakeholders using a variety of information retrieval systems including:

A. Project meetings

B. Fax

C. Electronic mail

D. Electronic databases

65. Assume that you have decided to use mitigation as a risk-response technique. This means:

A. You are shifting consequences of a risk to another party.

B. You are reducing the probability and/or consequences of an adverse risk event to an acceptable threshold.

C. You now need to establish a contingency allowance.

D. Your next step should be to prepare a fallback plan.

66. The traditional or functional organizational form has the disadvantage of:

A. Poorly established vertical communications channels

B. No single focal point for clients/sponsors

C. Ineffective technical control

D. Inflexible use of personnel

67. Which of the following is not a basic element of contracts?

   A. Consideration

   B. Mutual agreement

   C. Level of effort

   D. Legal purpose

68. Taking action to increase the effectiveness and efficiency of the project to provide added benefits to the stakeholders is the purpose of:

   A. Quality planning

   B. Inspections

   C. Quality audits

   D. Quality improvement efforts

69. During the procurement planning process, it is important to assess the current project boundaries. This can be done by reviewing the:

   A. Results of the make-or-buy analysis

   B. Product description

   C. Scope statement

   D. Constraints and assumptions

70. In project communications management, in order to ensure that the information needs of various stakeholders are met, you should:

   A. Prepare a stakeholder analysis.

   B. Establish an information distribution system.

   C. Assess communications skills.

   D. Evaluate available communications technologies.

71. Assigning resources in an attempt to find the shortest project schedule consistent with *fixed* resource limits is called:

   A. Resource allocation

   B. Resource partitioning

   C. Resource leveling

   D. Resource quantification

72. The process of assessing the impact and exposure of identified risks is known as:

   A. Risk-management planning

   B. Risk-response planning

   C. Qualitative risk analysis

   D. Quantitative risk analysis

73. An advantage of the analogous cost-estimating technique is:

   A. It provides greater accuracy than parametric estimating.

   B. Historical information is not required.

   C. Expert judgment is never needed.

   D. Lower costs are involved in its use than with definitive estimates.

74. Action taken to bring a defective or nonconforming item in compliance with requirements or specifications is the purpose of:

    A. Rework

    B. Control charts

    C. Audits

    D. Process adjustments

75. If you want to describe where the project now stands, you should:

    A. Prepare an estimate to complete.

    B. Prepare an earned value analysis.

    C. Prepare a status report.

    D. Prepare a progress report.

76. One purpose of risk control is to:

    A. See if assumptions are still valid.

    B. Determine whether risk-response actions are as effective as expected.

    C. Assess whether a risk trigger has occurred.

    D. Take corrective action.

77. In source selection, a weighting system may be used for all but one of the following purposes:

    A. To rank-order all proposals to establish a negotiating sequence

    B. To select a single source who will be asked to sign a standard contract

    C. To establish minimum requirements of performance for one or more evaluation criteria

    D. To quantify qualitative data to minimize the effect of personal prejudice on source selection

78. The overall intentions and directions of an organization with regard to quality is the purpose of:

    A. The total quality management movement

    B. The quality assurance process

    C. The quality planning process

    D. The organization's quality policy

79. The project communications management plan should:

    A. State communications skills to use.

    B. Describe methods used to gather and store information.

    C. Provide information to stakeholders as to how resources are being used to meet project objectives.

    D. Describe relationships between the organization and stakeholders.

80. During a project review meeting, you discover that the planned value is $400,000, the actual costs are $325,000, and the earned value is $300,000. You can therefore conclude that:

A. The project is behind schedule and overrunning costs.

B. The project is ahead of schedule, but costs are higher than budgeted.

C. The project is behind schedule with costs under control.

D. The project is on schedule, but costs are higher than budgeted.

▶ **Answer Key**

Using the answer key, score yourself and fill in the tables in Exhibit 1. Give yourself 10 points for each correct answer and no points for an incorrect answer. After you fill in the tables in Exhibit 1, continue on for an interpretation of your results.

1. D	**21.** C	**41.** D	**61.** C
2. B	**22.** D	**42.** A	**62.** A
3. A	**23.** B	**43.** B	**63.** A
4. A	**24.** D	**44.** A	**64.** D
5. A	**25.** B	**45.** C	**65.** B
6. B	**26.** C	**46.** D	**66.** B
7. B	**27.** B	**47.** D	**67.** C
8. C	**28.** A	**48.** D	**68.** D
9. A	**29.** D	**49.** A	**69.** C
10. D	**30.** B	**50.** A	**70.** A
11. C	**31.** C	**51.** B	**71.** A
12. A	**32.** A	**52.** B	**72.** C
13. B	**33.** A	**53.** C	**73.** D
14. C	**34.** B	**54.** A	**74.** A
15. D	**35.** B	**55.** A	**75.** C
16. A	**36.** D	**56.** B	**76.** D
17. C	**37.** D	**57.** D	**77.** C
18. A	**38.** C	**58.** B	**78.** D
19. D	**39.** A	**59.** A	**79.** B
20. B	**40.** B	**60.** D	**80.** A

■ Exhibit 1

Put the points in the space provided by each question, and then total each category.

Scope Management	Time Management	Cost Management	Human Resources Management
1. _____	2. _____	4. _____	5. _____
16. _____	17. _____	10. _____	9. _____
21. _____	24. _____	18. _____	15. _____
27. _____	31. _____	26. _____	19. _____
32. _____	33. _____	37. _____	28. _____
38. _____	48. _____	44. _____	46. _____
41. _____	51. _____	50. _____	52. _____
45. _____	58. _____	61. _____	55. _____
47. _____	63. _____	73. _____	57. _____
60. _____	71. _____	80. _____	66. _____
Total _____	Total _____	Total _____	Total _____

Procurement Management	Quality Management	Risk Management	Communication Management
6. _____	8. _____	7. _____	3. _____
13. _____	12. _____	14. _____	11. _____
23. _____	22. _____	25. _____	20. _____
34. _____	36. _____	29. _____	30. _____
40. _____	43. _____	39. _____	35. _____
49. _____	54. _____	42. _____	56. _____
59. _____	62. _____	53. _____	64. _____
67. _____	68. _____	65. _____	70. _____
69. _____	74. _____	72. _____	75. _____
77. _____	78. _____	76. _____	79 _____
Total _____	Total _____	Total _____	Total _____

Category	Points
Scope Management	
Time Management	
Cost Management	
Human Resources Management	
Procurement Management	
Quality Management	
Risk Management	
Communications Management	
Total	

▶ Explanation of Points for Level 1

If you received a score of 60 or more points in each of the eight categories, then you have a reasonable knowledge of the basic principles of project management. If you received a score of 60 or more in all but one or two of the categories, it's possible that you and your organization still possess all the knowledge you need of basic principles but that one or two of the categories do not apply directly to your circumstances. For example, if most of your projects are internal to your organization, procurement management may not be applicable. Also, for internal projects, companies often do not need the rigorous cost-control systems that would be found in project-driven organizations. Eventually, however, specialized training in these deficient areas will be needed.

If your score is less than 60 in any category, a deficiency exists. For scores less than 30 in any category, rigorous training programs on basic principles appear necessary. The organization appears highly immature in project management.

A total score on all categories of 600 or more would indicate that the organization appears well positioned to begin work on Level 2 of the PMMM. If your organization as a whole scores less than 600 points, there may exist pockets of project management. Each pocket may be at a different level of knowledge. Project-driven pockets generally possess more project management knowledge than non–project-driven pockets.

This assessment instrument can be used to measure either an individual's knowledge or an organization's knowledge. To assess organizational knowledge accurately, however, care must be taken in determining the proper cross-section of participants to be tested.

▶ Opportunities for Customizing Level 1

Level 1 assessment questions can be customized based on the knowledge areas most commonly used in the company. At present, there are 10 knowledge areas in the *PMBOK® Guide*, but only 8 areas were tested on here. Some companies prefer customization to expand on the knowledge areas that are of a greater concern to the firm. For example, the questions related to quality management may be more appropriate to a firm that has manufacturing capability and uses statistical process-control charts. If this is not the case, then the quality management questions could be replaced with questions related to integration management or stakeholder management.

Some of the questions could be replaced with questions related to other knowledge areas or even processes that are commonly used by the firm but are not part of the *PMBOK® Guide*. Level 1 is probably the easiest level for customization.

CHAPTER SIX

Level 2: Common Processes

▶ Introduction

Learning the basics of project management, and even having several employees certified as project management professionals (PMP® credential holders), does not guarantee that project management is being used in your organization. Even if it is being used, it may not be used effectively. Level 2 is the stage where an organization makes a concerted effort to use project management and to develop processes and methodologies to support its effective use.

In Level 2, the organization realizes that common methodologies and processes are needed such that managerial success on one project can be repeated on other projects. Also apparent in this level is the fact that certain behavioral expectations of organizational personnel are necessary for the repetitive execution of the methodology. These are the characteristics of Level 2, as shown in Figure 6.1:

- Tangible benefits of using project management must become apparent. The most common benefits include lower cost, shortened schedules, no sacrifice of scope or quality, and the potential for a higher degree of customer satisfaction.

- Project management must be supported throughout all levels of the organization, including the senior levels. It is possible that changes to the corporate culture may be necessary, thus mandating executive support.

- A continuous stream of successfully managed projects requires methodologies and processes that can be used over and over again. This requires an organizational commitment.

- Managing projects within scope and time is only part of the effort. The projects must also be completed within cost, and this may mandate changes to the cost accounting system.

- The final characteristic of Level 2 is the development of a project management curriculum rather than just a project management course. This is often seen as proof of the organization's firm commitment to project management.

PMP is a registered mark of the Project Management Institute, Inc.

> **Common processes**
> - Recognition of the benefits of project management
> - Organizational support at all levels
> - Recognition of the need for processes/methodologies
> - Recognition of the need for cost control
> - Development of a project management training curriculum

Figure 6.1 Characteristics of Level 2.

These bullets are the outputs of the common processes. In other words, do you have common processes to facilitate repeatability? It should be noted that many of the benefits of common processes many not be clearly visible. The benefits may be intangible at first, and then become visible benefits later.

▶ Life Cycles for Level 2

Common processes require a good process-definition effort accompanied by the necessary organizational behavior needed for the execution of the processes. Level 2, common processes, can be broken down into five life-cycle phases, as shown in Figure 6.2. These life-cycle phases are actually subphases or steps within Level 2 of the project management maturity model (PMMM) in order to develop common processes. The first life-cycle phase of Level 2 is the *embryonic phase*, which is where the organization recognizes that project management can benefit the organization. The embryonic phase includes:

- Recognizing the need for project management
- Recognizing the potential benefits of project management
- Recognizing the applications of project management to the various parts of the business
- Recognizing some of the changes necessary to implement project management

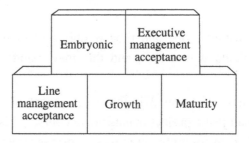

Figure 6.2 Life-cycle phases for Level 2 of project management maturity.

Companies do not generally promote the acceptance of project management unless they understand a sound basis for wanting project management. The six most common driving forces for project management are as follows:

- *Capital projects:* High-dollar-value capital projects require effective planning and scheduling. Without project management, ineffective use of manufacturing resources may occur.

- *Customer expectations:* Customers have the right to expect the contractor to manage the customer's work requirements efficiently and effectively.

- *Internal competitiveness:* Executives want employees to focus on external competition rather than internal competition, power struggles, and gamesmanship.

- *Executive understanding:* Although it's uncommon, executives can drive the acceptance of project management from the top of the organization down to the bottom.

- *New-product development:* Executives want a methodology in place that provides a high likelihood that R&D projects will be completed successfully, in a timely manner, and within reasonable cost.

- *Efficiency and effectiveness:* Executives want the organization to be highly competitive.

In theory, most companies have one and only one driving force. While we've just discussed six different driving forces, in practice they combine to give one, and only one: survival. This is shown in Figure 6.3. Once executives recognize that project management may be needed for survival, changes occur quickly.

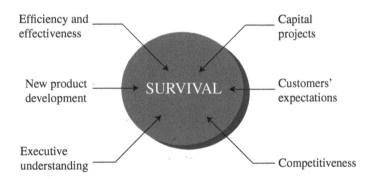

Figure 6.3 The components of survival.

What is unfortunate about the embryonic phase is that the recognition of benefits and applications may be seen first by lower and middle levels of management. Senior management must then be "sold" on the concept of project management. This leads us to the second life-cycle phase: *executive management acceptance.* Included in the executive management acceptance phase are the following:

- Visible executive support

- Executive understanding of project management

- Project sponsorship and/or governance

- Willingness to change the way the company does business

The third life-cycle phase of Level 2 is *line management acceptance*. This includes:

■ Visible line-management support

■ Line management commitment to project management

■ Line management education

■ Release of functional employees for project management training programs

It is highly unlikely that line managers will provide support for project management unless they also see visible executive support.

The fourth life-cycle phase of Level 2 is the *growth* phase. This is the critical phase. Although some of the effort in this phase can be accomplished in parallel with the first three life-cycle phases of Level 2, the completion of this phase is predicated upon the completion of the first three life-cycle phases. The growth phase is the beginning of the creation of the project management process. Included in this phase are:

■ Development of company project management life cycles

■ Development of a project management methodology

■ A commitment to effective planning

■ Minimization of scope changes (i.e., of creeping scope)

■ Selection of project management software to support the methodology

Unfortunately, companies often develop several types of methodologies for each type of project within the organization. This becomes an inefficient use of resources, although it can function as a good learning experience for the company.

The fifth life-cycle phase of Level 2 is the *initial maturity phase*. Included in this phase are:

■ The development of a management cost/schedule control system

■ Integration of schedule and cost control

■ Development of an ongoing educational curriculum to support project management and enhance individual skills

Many companies never fully complete this life-cycle phase because the organization is resistant to project cost control, otherwise known as *horizontal accounting*. Line managers dislike horizontal accounting because it clearly identifies which line managers provide good estimates for projects and which do not. Executives resist horizontal accounting because the executives want to establish a budget and schedule long before a project plan is created.

▶ Roadblocks

Figure 6.4 illustrates the most common roadblocks that prevent an organization from completing Level 2. Based on the strength and longevity of the corporate culture, there could be strong resistance to change. The argument is always, "What we already have works well." The resistance to change stems from the fear that support

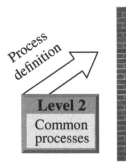

● Resistance to a new
   methodology
● What we already have
   works well
● Belief that a
   methodology needs
   rigid policies and
   procedures
● Resistance to
   horizontal accounting

Figure 6.4 Roadblocks to completion of Level 2.

for a new methodology will result in a shift in established power and authority relationships.

Another area of resistance is due to the misbelief that a new methodology must be accompanied by rigid policies and procedures, thus once again causing potential changes to the power and authority structure. The final roadblock comes from the fear that horizontal accounting will bring to the surface problems that people would prefer to keep hidden, such as poor estimating ability.

► Advancement Criteria

Four key actions are required to complete Level 2 and advance to Level 3:

■ Develop a culture that supports both the behavioral and quantitative sides of project management.

■ Recognize both the driving forces/need for project management and the benefits that can be achieved in both the short term and the long term.

■ Develop a project management process/methodology such that the desired benefits can be achieved on a repetitive basis.

■ Develop an ongoing, all-employee project management curriculum such that project management benefits can be sustained and improved upon for the long term.

► Risk

The successful completion of Level 2 usually occurs with a medium degree of difficulty. The time to complete Level 2 is usually six months to two years, based on such factors as the following:

■ Type of company (project-driven versus non–project-driven)

■ Visibility of executive support

■ Strength of the corporate culture

- Resistance to change
- Speed with which a good, workable methodology can be developed
- Existence of an executive-level champion to drive the development of the project management methodology
- Using a PMO to lead the effort
- Speed with which project management benefits can be realized

The risk in this level can be overcome through strong, visible executive support.

▶ Overlapping Levels

Level 2 can and does overlap Level 1. There is no reason you must wait for a multitude of people to be trained in project management before you begin the development of processes and methodologies. Also, the earlier the company begins developing processes and methodologies, the earlier those processes and methodologies can be included as part of the training. One company conducted a three-day course on the principles of project management. A fourth day was spent covering the company's processes and methodologies for project management. Thus, the employees could see clearly how the processes/methodologies utilized the basic concepts of project management.

▶ Assessment Instrument for Level 2

Level 2, common processes, is the process-definition level. Level 2 can be fulfilled by recognizing the different life-cycle phases of Level 2.

The following 20 questions explore how mature you believe your organization to be in regard to Level 2 and the accompanying life-cycle phases of Level 2. Beside each question, circle the number that corresponds to your opinion. In the following example, your choice would have been "slightly agree":

−3 Strongly disagree

−2 Disagree

−1 Slightly disagree

0 No opinion

(+1) Slightly agree

+2 Agree

+3 Strongly agree

Example: (−3, −2, −1, 0, (+1), +2, +3)

The row of numbers from −3 to +3 will be used later for evaluating the results. After answering question 20, you will grade the exercise.

► Questions

The following 20 questions involve Level 2 maturity. The questions look at both repeatability of processes and actions necessary to support continuous development of processes. Please answer each question as honestly as possible. Circle the answer you feel is correct.

1. My company recognizes the need for project management. This need is recognized at all levels of management, including senior management.

   (−3 −2 −1 0 +1 +2 +3)

2. My company has a system in place to manage both cost and schedule. The system requires charge numbers and cost account codes. The system reports variances from planned targets.

   (−3 −2 −1 0 +1 +2 +3)

3. My company has recognized the benefits that are possible from implementing project management. These benefits have been recognized at all levels of management, including senior management.

   (−3 −2 −1 0 +1 +2 +3)

4. My company (or division) has a well-definable project management methodology using life-cycle phases.

   (−3 −2 −1 0 +1 +2 +3)

5. Our executives visibly support project management through executive presentations and correspondence, and by occasionally attending project team meetings/briefings.

   (−3 −2 −1 0 +1 +2 +3)

6. My company is committed to quality up-front planning. We try to do the best we can at planning.

   (−3 −2 −1 0 +1 +2 +3)

7. Our lower- and middle-level line managers totally and visibly support the project management process.

   (−3 −2 −1 0 +1 +2 +3)

8. My company is doing everything possible to minimize creeping scope (i.e., scope changes) on our projects.

   (−3 −2 −1 0 +1 +2 +3)

9. Our line managers are committed not only to project management but also to the promises made to project managers for deliverables.

   (−3 −2 −1 0 +1 +2 +3)

10. The executives in my organization have a good understanding of the principles of project management.

    (−3 −2 −1 0 +1 +2 +3)

11. My company has selected one or more project management software packages to be used as the project tracking system.

    (−3 −2 −1 0 +1 +2 +3)

12. Our lower- and middle-level line managers have been trained and educated in project management.

    (−3   −2   −1   0   +1   +2   +3)

13. Our executives both understand project sponsorship and serve as project sponsors on selected projects.

    (−3   −2   −1   0   +1   +2   +3)

14. Our executives have recognized or identified the applications of project management to various parts of our business.

    (−3   −2   −1   0   +1   +2   +3)

15. My company has successfully integrated cost and schedule control for both managing projects and reporting status.

    (−3   −2   −1   0   +1   +2   +3)

16. My company has developed a project management curriculum (i.e., more than one or two courses) to enhance the project management skills of our employees.

    (−3   −2   −1   0   +1   +2   +3)

17. Our executives have recognized what must be done in order to achieve maturity in project management.

    (−3   −2   −1   0   +1   +2   +3)

18. My company views and treats project management as a profession rather than a part-time assignment.

    (−3   −2   −1   0   +1   +2   +3)

19. Our lower- and middle-level line managers are willing to release their employees for project management training.

    (−3   −2   −1   0   +1   +2   +3)

20. Our executives have demonstrated a willingness to change our way of doing business in order to mature in project management.

    (−3   −2   −1   0   +1   +2   +3)

Now turn to Exhibit 2 and grade your answers.

■ Exhibit 2

Each response you circled in questions 1–20 had a column value between −3 and +3. In the appropriate spaces below, place the circled value (between −3 and +3) beside each question.

Embryonic

1. _____

3. _____

14. _____

17. _____

TOTAL _____

Executive Management

5. _____

10. _____

13. _____

20. _____

TOTAL _____

Line Management

7. _____

9. _____

12. _____

19. _____

TOTAL _____

Growth

4. _____

6. _____

8. _____

11. _____

TOTAL _____

Maturity

2. _____

15. _____

16. _____

18. _____

TOTAL _____

Transpose your total score in each category to the following table by placing an X in the appropriate area.

Points Life-cycle phases	−12	−10	−8	−6	−4	−2	0	+2	+4	+6	+8	+10	+12
Maturity													
Growth													
Line Management													
Executive													
Embryonic													

▶ Explanation of Points for Level 2

High scores (usually +6 or greater) for a life-cycle phase indicate that these evolutionary phases of early maturity have been achieved or at least you are now in this phase. Phases with very low numbers have not been achieved yet.

Consider the following scores:

Embryonic	+ 8
Executive	+10
Line Management	+ 8
Growth	+ 3
Maturity	4

This result indicates that you have probably completed the first three stages and are now entering the growth phase. Keep in mind that the answers are not always this simple because companies can achieve portions of one stage in parallel with portions of a second or third phase.

▶ Opportunities for Customizing Level 2

Level 2 focuses heavily on culture. Many of the questions can be changed to include how much executive and line management support exists, whether you use virtual teams and the support for virtual teams, how well functional organizations work with one another, and stakeholder relationship management issues. If the customization is done carefully, the grading system may not need to be changed.

CHAPTER SEVEN

Level 3: Singular Methodology

▶ Introduction

Level 3 is the level at which the organization recognizes that synergism and process control can best be achieved through the development of a singular methodology rather than by using multiple methodologies for the same group or domain of projects. There may exist a separate methodology for new product development and another methodology for information systems. The goal, however, should be to determine the minimum number of domains or groups and have one methodology for each.

When companies first start out in project management, the goal is a singular methodology that allows the organization (especially at the senior management levels) to maintain some degree of standardization and control over projects. As companies develop some degree of maturity, the singular approach becomes a flexible approach where each project manager can create their own methodology from the existing forms, guidelines, templates, and checklists to satisfy a client. As organizations progress in the maturity process, flexible methodologies and frameworks, such as with agile and Scrum, appear. However, the assumption made at this level is that the organization may just be starting out in project management. Modifications can be made to this level to account for flexible methodologies.

At this level, the organization is totally committed to the concept of project management. The characteristics of Level 3, as shown in Figure 7.1, are as follows:

- *Integrated processes:* The organization recognizes that multiple processes can be streamlined into a single integrated process encompassing all other processes. (However, not all companies have the luxury of using a single methodology nor do they desire to do so.)

- *Cultural support:* Integrated processes create a singular methodology. It is through this singular methodology that exceptional benefits are achieved. The execution of the methodology is through the corporate culture, which now wholeheartedly supports the project management approach. The culture becomes a cooperative culture.

- *Management support:* At this level, project management support permeates the organization throughout all layers of management. The support is visible. Each layer or level of management understands its role and the support needed to make the singular methodology work.

- *Informal project management:* With management support and a cooperative culture, the singular methodology is based on guidelines and checklists, rather than the expensive development of rigid policies and procedures. Paperwork is minimized.

- *Training and education:* With strong cultural support, the organization realizes financial benefits from project management education. The benefits can be described quantitatively and qualitatively.

- *Behavioral excellence:* The organization recognizes the behavioral differences between project management and line management. Behavioral training programs are developed to enhance project management skills.

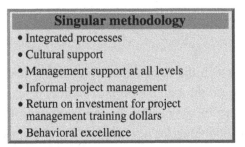

Singular methodology
- Integrated processes
- Cultural support
- Management support at all levels
- Informal project management
- Return on investment for project management training dollars
- Behavioral excellence

Figure 7.1 Characteristics of Level 3.

These six characteristics formulate the *hexagon of excellence*, as shown in Figure 7.2. These six areas differentiate those companies excellent in project management from those with average skills in project management. Each of the six areas is discussed next.

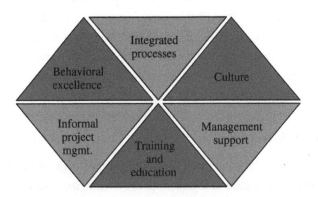

Figure 7.2 The hexagon of excellence.

▶ Integrated Processes

Companies that are relatively immature in project management have multiple processes in place. Figure 7.3 shows the three most common of these separate processes.

Concurrent engineering, for those unfamiliar with the term, is similar to fast-tracking a project where activities are overlapping, in order to accelerate the completion date. Why, however, would a company want its processes, its facilities, its resources in general, to be totally uncoupled? The first two processes to be integrated, once an organization understands the advantages, are usually project management and total quality management (TQM) or other quality improvement techniques such as Six Sigma. After all, employees trained in the principles of TQM will realize the similarities between the two processes. All the winners of the prestigious Malcolm Baldrige National Quality Award appear to have excellent project management systems in place.

When organizations begin to realize the importance of a singular methodology, project management becomes integrated with quality improvement practices and concurrent engineering to formulate a singular methodology. This integration is shown in Figure 7.4. As companies begin to climb the ladder toward excellence in project management, the initial singular methodology is further enhanced to include risk management and change management, as shown in Figure 7.5. Risks generally require scope changes, which, in turn, create additional risks. Creating a singular, integrated methodology that encompasses all other methodologies leads to organizational efficiency and effectiveness.

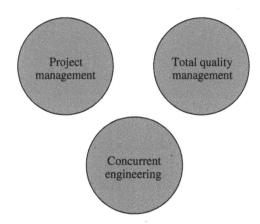

Figure 7.3   Totally uncoupled processes.

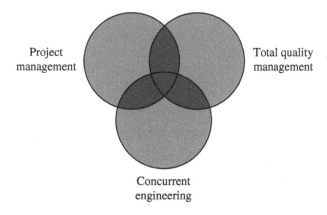

Figure 7.4   Totally integrated processes.

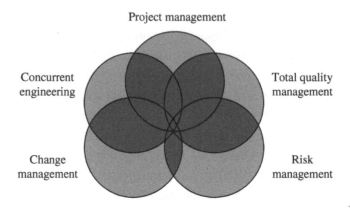

Figure 7.5 Integrated processes for the twenty-first century.

As a final note, not all companies have the luxury or ability to utilize one and only one methodology for the execution of all projects. Some companies readily admit that two or more methodologies might be needed, such as one for new product development and a second for information systems projects. Therefore, it may be more appropriate to recognize this level as an attempt to get everyone to agree to a singular methodology, if possible.

▶ Culture

Project management methodologies must not simply be pieces of paper. The pieces of paper must be converted into a world-class methodology by the way in which the corporate culture executes the methodology. Companies excellent in project management have cooperative cultures where the entire organization supports the singular methodology. Organizational resistance is at a minimum, and everyone pitches in during times of trouble.

Cultural transformation is never easy. From a project management perspective, common purposes for a cultural transformation include corporate vision, corporate goals, and the creation of a singular methodology for project management. Although there are a multitude of cultural issues, the four most common fears affecting project management are:

- The embedded fear of change
- The fear of having to create a new social group
- The fear of a change in work habits or comfort zone
- The uncertain impact on the wage and salary program

The hierarchy of fears can differ from company to company, but strategies must be put in place to overcome these fears or else Level 3 completion could be prolonged. Typical ways to overcome these fears are shown in Figure 7.6.

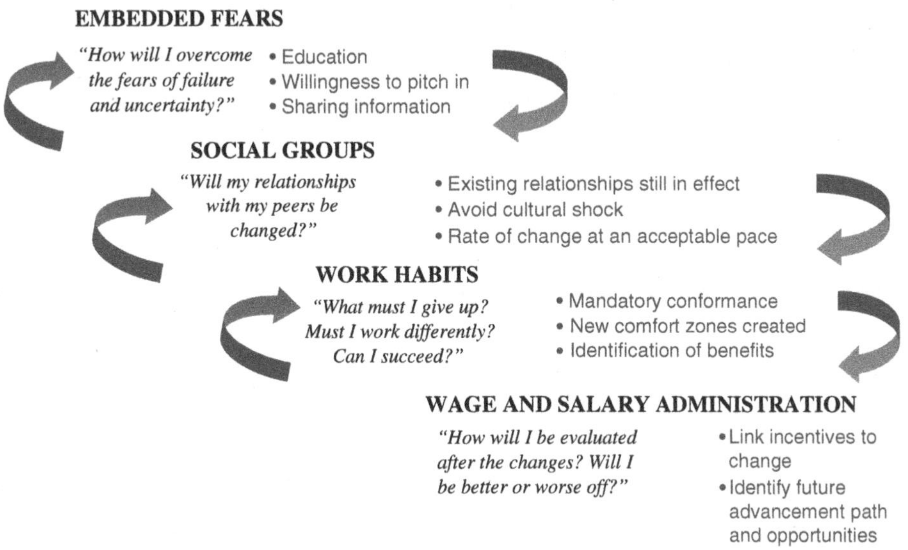

EMBEDDED FEARS

"How will I overcome the fears of failure and uncertainty?"
- Education
- Willingness to pitch in
- Sharing information

SOCIAL GROUPS

"Will my relationships with my peers be changed?"
- Existing relationships still in effect
- Avoid cultural shock
- Rate of change at an acceptable pace

WORK HABITS

"What must I give up? Must I work differently? Can I succeed?"
- Mandatory conformance
- New comfort zones created
- Identification of benefits

WAGE AND SALARY ADMINISTRATION

"How will I be evaluated after the changes? Will I be better or worse off?"
- Link incentives to change
- Identify future advancement path and opportunities

Figure 7.6 Ways to overcome resistance to change.

▶ Management Support

Cooperative cultures require effective management support at all levels. During the execution of the project management methodology, the interface between project management and line management is critical. Effective relationships with line management are based on these factors:

- Project managers and line managers share accountability for the successful completion of a project. Line managers must keep their promises to project managers.

- Project managers negotiate with line managers for the accomplishment of deliverables rather than for specific talent. Project managers can request specific talent, but the final decision for staffing belongs to the line manager.

- Line managers trust their employees enough to empower those employees to make decisions related to their specific functional area without continuously having to run back to their line manager.

- If a line manager is unable to keep a promise they made to a project, then the project manager must do everything possible to help the line manager develop alternative plans.

The relationship between project management and senior management is equally important. A good relationship with executive management, specifically the executive sponsor, includes these factors:

- The project manager is empowered to make project-related decisions. This is done through decentralization of authority and decision-making.

- The sponsor is briefed periodically while maintaining a hands-off, but available, position.

- The project manager (and other project personnel) are encouraged to present recommendations and alternatives rather than just problems.

- Exactly what needs to be included in a meaningful executive status report has been formulated.

- A policy is in place calling for periodic, but not excessively frequent, briefings.

▶ Informal Project Management

With informal project management, the organization recognizes the high cost of paperwork. Informal project management does not eliminate paperwork. Instead, paperwork requirements are reduced to the minimum acceptable levels. For this to work effectively, the organization must experience effective communications, cooperation, trust, and teamwork. These four elements are critical components of a cooperative culture.

As trust develops, project sponsorship may be pushed down from the executive levels to middle management. The project manager no longer wears multiple hats (i.e., being a project manager and line manager at the same time), but functions as a dedicated project manager.

The development of project management methodologies at Level 2 are based on rigid policies and procedures. But at Level 3, with a singular methodology based more on informal project management, methodologies are written in the format of general guidelines and checklists. This drastically lowers methodology execution cost and execution time.

Dashboard reporting systems have allowed companies to provide additional information without words. As an example, one company uses a "traffic light" beside each work breakdown structure (WBS) work package in the status report. The traffic light is either red, yellow, or green, based on the following definitions:

- *Red:* A problem exists that may affect time, cost, scope, or quality. Sponsor involvement is necessary.

- *Yellow:* This is a caution. A potential problem may exist. The sponsor is informed, but no action by the sponsor is necessary right now.

- *Green:* Work is progressing as planned. Sponsor involvement is not necessary.

▶ Training and Education

At Level 3, there is a recognition that there exists a return on investment for training dollars. The benefits, or return on investment, can be measured quantitatively and qualitatively. Quantitative results include:

- Shorter product development time

- Faster, higher-quality decisions

- Lower costs

- Higher profit margins

- Fewer people needed

- Reduction in paperwork
- Improved quality and reliability
- Lower turnover of personnel
- Quicker "best practices" implementation

Qualitative results include:

- Better visibility and focus on results
- Better coordination
- Higher morale
- Accelerated development of managers
- Better control
- Better customer relations
- Better support from the functional areas
- Fewer conflicts requiring senior management involvement

Project management training and education is an investment and, as such, senior management wishes to know when the added profits will materialize. This can best be explained from Figure 7.7. Initially, a substantial cost may be incurred during Level 2 and the beginning of Level 3. But as the culture develops and informal project management matures, the cost of project management diminishes to a pegged level while the additional profits grow. The question mark in Figure 7.7 generally occurs during Level 3, which is usually about two to five years after the organization has made a firm commitment to project management.

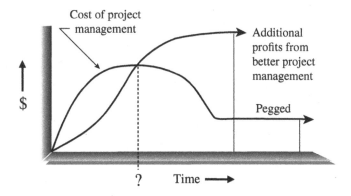

Figure 7.7 Project management costs versus benefits.

A question normally asked by executives is, "How do we know if we are in Level 3 of the project management maturity model (the PMMM)?" The answer is, by the number of conflicts coming up to the senior levels of management for resolution.

By Level 3, executives have realized that the speed at which the benefits can be achieved can be accelerated through proper training and education. Therefore, the training and education in Level 3 does not consist merely of a few random courses. Instead, as discussed in the advancement criteria for completing Level 2 and moving up to Level 3, the company develops a project management curriculum. This will encompass a *core competency model* for the basic and advanced skills that a project manager should possess. Training is conducted to support the core competency skills.

▶ Behavioral Excellence

Behavioral excellence occurs when the organization recognizes the differences between project management and line management, and the fact that a completely different set of training courses is required to support sustained project management growth. Emphasis is placed on:

- Motivation in project management
- Creation of outstanding project leaders
- Characteristics of productive teams
- Characteristics of productive organizations
- Sound and effective project management practices

People are often under the misapprehension that achieving Level 3 in the PMMM will deliver 100 percent successful projects. This is not true. Successful implementation of project management does not guarantee that your projects will be successful. Instead, it does guarantee that your projects will be managed effectively, thus improving your chances of success. From Figure 7.8 you can see that, during Level 3 of the PMMM, the number of project successes increases. However, even though the number of successes increases and comes to dramatically exceed the number of failures, failure still exists. Project management does not circumvent the problem of unrealistic objectives or targets, unforeseen acts of God, poor decision-making by people with authority, and economic upheaval. Any company that has a 100 percent project success rate is not working on enough projects. No risk is being taken. Also, any executive sponsor or project manager who always makes the right decision is probably not making enough decisions.

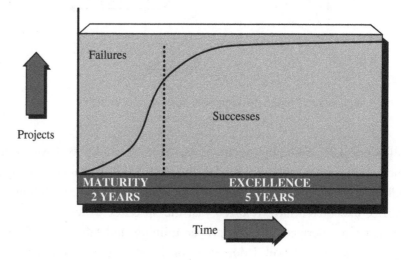

Figure 7.8 Growth in successes.

► Roadblocks

Figure 7.9 shows the key roadblocks that prevent an organization from completing Level 3. They include:

- Senior management may have the attitude, "Don't fix it if it isn't broken" and feel that the organization should continue to use the processes now in place.

- There will always exist initial resistance to a singular methodology, for fear that it will be accompanied by shifts in the balance of power.

- Line managers may resist accepting accountability for the promises made to the projects. Shared accountability is often viewed as a high risk for line managers.

- Organizations with strong, fragmented corporate cultures often resist being converted over to a single, cooperative culture.

- Some organizations thrive on the belief that what is not on paper has not been said. Overemphasis on documentation is a bad habit that is hard to break.

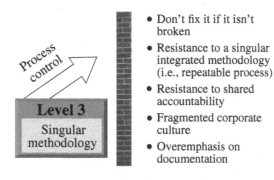

Figure 7.9 Roadblocks to completion of Level 3.

► Advancement Criteria

Certain key actions are needed, to advance from Level 3 to Level 4. These key actions are as follows:

- Integrate all related processes into a single methodology with demonstrated successful execution.

- Encourage the corporate-wide acceptance of a culture that supports informal project management and multiple-boss reporting.

- Develop support for shared accountability.

► Risk

The successful completion of Level 3 is accompanied by a high degree of difficulty if the culture of the company must change. Culture shock may result especially if the change is seen as a radical modification that removes people from their comfort zone. The time period to complete Level 3 may be measured in years, based on such factors as:

- The speed at which the culture will change
- The acceptance of informal project management
- The acceptance of a singular methodology

The greatest degree of risk in project management is attributed to the corporate culture. Poorly designed methodologies can convert a good, cooperative culture into a combative culture.

If an organization develops a singular methodology (or, if necessary, one methodology for each domain), then the organization should strive for corporate-wide acceptance of that methodology for each domain. If the methodology is accepted and used only in pockets of interest, then a fragmented culture will occur. Fragmented cultures generally do not allow the organization to maximize the benefits of project management.

► Overlapping Levels

Generally speaking, Levels 2 and 3 do not overlap. Once a company recognizes the true benefits of project management and the need for a singular methodology, the organization stops developing individual processes and focuses on what's best for the whole.

Allowing individual processes to continue without any integration into a singular methodology gives employees a viable excuse to resist change. Employees must be encouraged to make decisions that are in the best interest of the entire company rather than in the best interest of their own department.

► Assessment Instrument for Level 3

The following 42 multiple-choice questions will allow you to compare your organization against other companies with regard to the Level 3 hexagon of excellence. After you complete question 42, a grading system is provided. You can then compare your organization to some of the best that have achieved Level 3 maturity.

Please pick one and only one answer per question. A worksheet and answer key follow the exercise.

► Questions

1. My company *actively* uses the following processes:
   A. Total quality management (TQM) or other quality initiatives only
   B. Concurrent engineering (shortening deliverable development time) only

C. TQM or other quality initiatives and concurrent engineering only

D. Risk management only

E. Risk management and concurrent engineering only

F. Risk management, concurrent engineering, and TQM or other quality initiatives

2. On what percentage of your projects do you use the principles of total quality management?

   A. 0%

   B. 5–10%

   C. 10–25%

   D. 25–50%

   E. 50–75%

   F. 75–100%

3. On what percentage of your projects do you use the principles of risk management?

   A. 0%

   B. 5–10%

   C. 10–25%

   D. 25–50%

   E. 50–75%

   F. 75–100%

4. On what percentage of your projects do you try to compress product/deliverable schedules by performing work in parallel rather than in series?

   A. 0%

   B. 5–10%

   C. 10–25%

   D. 25–50%

   E. 50–75%

   F. 75–100%

5. My company's risk management process is based on:

   A. We do not use risk management.

   B. Financial risks only

   C. Technical risks only

   D. Scheduling risks only

   E. A combination of financial, technical, and scheduling risks based on the project

6. The risk management methodology in my company is:

   A. Nonexistent

   B. More informal than formal

C. Based on a structured methodology supported by policies and procedures

D. Based on a structured methodology supported by policies, procedures, and standardized forms to be completed

7. How many different project management methodologies exist in your organization (i.e., do you consider a systems development methodology for MIS projects different than a product development project management methodology)?

    A. We have no methodologies.

    B. 1

    C. 2–3

    D. 4–5

    E. More than 5

8. With regard to benchmarking:

    A. My company has never tried to use benchmarking.

    B. My company has performed benchmarking and implemented changes, but not for project management.

    C. My company has performed project management benchmarking, but no changes were made.

    D. My company has performed project management benchmarking, and changes were made.

9. My company's corporate culture is best described by the concept of:

    A. Single-boss reporting

    B. Multiple-boss reporting

    C. Dedicated teams without empowerment

    D. Nondedicated teams without empowerment

    E. Dedicated teams with empowerment

    F. Nondedicated teams with empowerment

10. With regard to morals and ethics, my company believes that:

    A. The customer is always right.

    B. Decisions should be made in the following sequence: best interest of the customer first, then the company, then the employees.

    C. Decisions should be made in the following sequence: best interest of the company first, customer second, and the employees last.

    D. We have no such written policy or set of standards.

11. My company conducts internal training courses on:

    A. Morality and ethics within the company

    B. Morality and ethics in dealing with customers

    C. Good business practices

   D. All of the above

   E. None of the above

   F. At least two of the first three

12. With regard to scope creep or scope changes, our culture:

   A. Discourages changes after project initiation

   B. Allows changes only up to a certain point in the project's life cycle using a formal change-control process

   C. Allows changes anywhere in the project life cycle using a formal change-control process

   D. Allows changes but without any formal control process

13. Our culture seems to be based on:

   A. Policies

   B. Procedures (including forms to be filled out)

   C. Policies and procedures

   D. Guidelines

   E. Policies, procedures, and guidelines

14. Cultures are either quantitative (policies, procedures, forms, and guidelines), behavioral, or a compromise. The culture in my company is probably _____ behavioral.

   A. 10–25%

   B. 25–50%

   C. 50–60%

   D. 60–75%

   E. Greater than 75%

15. Our organizational structure is:

   A. Traditional (i.e., it is predominantly vertical).

   B. A strong matrix (i.e., project manager provides most of the technical direction).

   C. A weak matrix (i.e., line managers provide most of the technical direction).

   D. We use colocated teams.

   E. I don't know what the structure is: management changes it on a daily basis.

16. When assigned as a project leader, our project managers obtain resources by:

   A. "Fighting" for the best people available

   B. Negotiating with line managers for the best people available

   C. Negotiating for deliverables rather than people

   D. Using senior management to help get the appropriate people

   E. Taking whatever they get, no questions asked

17. Our line managers:

    A. Accept total accountability for the work in their line.

    B. Ask the project managers to accept total accountability.

    C. Try to share accountability with the project managers.

    D. Hold the assigned employees accountable.

    E. We don't know the meaning of the word *accountability*; it is not part of our vocabulary.

18. In the culture within my company, the person most likely to be held accountable for the ultimate technical integrity of the final deliverable is/are:

    A. The assigned employees

    B. The project manager

    C. The line manager

    D. The project sponsor

    E. The whole team

19. In my company, the project manager's authority comes from:

    A. Within themselves—whatever they can get away with

    B. The immediate superior to the project manager

    C. Documented job descriptions

    D. Informally, through the project sponsor in the form of a project charter or appointment letter

20. After project go-ahead, our project sponsors tend to:

    A. Become invisible, even when needed

    B. Micromanage

    C. Expect summary-level briefings once a week

    D. Expect summary-level briefings once every two weeks

    E. Get involved only when a critical problem occurs or at the request of the project manager or line managers

21. What percentage of your projects have sponsors who are at the director level or above?

    A. 0–10%

    B. 10–25%

    C. 25–50%

    D. 50–75%

    E. More than 75%

22. My company offers approximately this many different *internal* training courses for the employees (courses that can be regarded as project-related):

    A. Fewer than 5

    B. 6–10

   C. 11–20

   D. 21–30

   E. More than 30

23. With regard to the previous answer, what percentage of the courses are more behavioral than quantitative?

   A. Less than 10%

   B. 10–25%

   C. 25–50%

   D. 50–75%

   E. More than 75%

24. My company believes that:

   A. Project management is a part-time job.

   B. Project management is a profession.

   C. Project management is a profession, and we should become certified as project management professionals, but at our own expense.

   D. Project management is a profession, and my company pays for our training to become certified as project management professionals.

   E. We have no project managers in my company.

25. My company believes that training should be:

   A. Performed at the request of employees

   B. Performed to satisfy a short-term need

   C. Performed to satisfy both long- and short-term needs

   D. Performed only if there exists a return on investment on training dollars

26. My company believes that the content of training courses is best determined by:

   A. The instructor

   B. The Human Resource department

   C. Management

   D. Employees who will receive the training

   E. Customization after an audit of the employees and managers

27. What percentage of the training courses in project management contain *documented* lessons-learned case studies from other projects within your company?

   A. None

   B. Less than 10%

   C. 10–25%

   D. 25–50%

   E. More than 50%

28. What percentage of the executives in your functional (not corporate) organization have attended training programs or executive briefings specifically designed to show executives what they can do to help project management mature?

A. None!

B. Less than 25%

C. 25–50%

D. 50–75%

E. More than 75%

29. In my company, employees are promoted to management because:

A. They are technical experts.

B. They demonstrate the administrative skills of a professional manager.

C. They know how to make sound business decisions.

D. They are at the top of their pay grade.

E. Our rank-and-file pool is over its numerical upper limits.

30. A report must be written and presented to the customer. Neglecting the cost to accumulate the information, the approximate cost per page for a typical report is:

A. I have no idea.

B. $100–200 per page

C. $200–500 per page

D. Greater than $500 per page

E. Free; exempt employees in my company prepare the reports at home on their own time.

31. The culture within my organization is best described as:

A. Informal project management based on trust, communication, and cooperation

B. Formality based on policies and procedures for everything

C. Project management that thrives on formal authority relationships

D. Executive meddling, which forces an overabundance of documentation

E. Nobody trusting the decisions of our project managers

32. What percentage of the project manager's time each week is spent preparing reports?

A. 5–10%

B. 10–20%

C. 20–40%

D. 40–60%

E. Greater than 60%

33. During project *planning*, most of our activities are accomplished using:

A. Policies

B. Procedures

C. Guidelines

D. Checklists

E. None of the above

34. The typical time duration for a project status-review meeting with senior management is:

   A. Less than 30 minutes

   B. 30–60 minutes

   C. 60–90 minutes

   D. 90 minutes–2 hours

   E. Greater than 2 hours

35. Our customers mandate that we manage our projects:

   A. Informally

   B. Formally, but with scope creep disallowed

   C. Formally, but with scope creep allowed

   D. It is our choice as long as the deliverables are met.

36. My company believes that *poor* employees:

   A. Should never be assigned to teams

   B. Once assigned to a team, are the responsibility of the project manager for supervision

   C. Once assigned to a team, are still the responsibility of their line manager for supervision

   D. Can be effective if assigned to the right team

   E. Should be promoted into management

37. Employees who are assigned to a project team (either full time or part time) have a performance evaluation conducted by:

   A. Their line manager only

   B. The project manager only

   C. Both the project and line managers

   D. Both the project and line managers, together with a review by the sponsor

38. The skills that will probably be most important for my company's project managers as we move into the twenty-first century are:

   A. Technical knowledge and leadership

   B. Risk management and knowledge of the business

   C. Integration skills and risk management

   D. Integration skills and knowledge of the business

   E. Communication skills and technical understanding

39. In my organization, the people assigned as project leaders are usually:

   A. First-line managers

   B. First- or second-line managers

   C. Any level of management

   D. Usually non-management employees

   E. Anyone in the company

40. The project managers in my organization have undergone at least some degree of training in:

   A. Feasibility studies

   B. Cost-benefit analyses

   C. Both A and B

   D. Our project managers are typically brought on board after project approval/award.

41. Our project managers are encouraged to:

   A. Take risks

   B. Take risks upon approval by senior management

   C. Take risks upon approval by project sponsors

   D. Avoid risks

42. Consider the following statement: Our project managers have a sincere interest in what happens to each team member *after* the project is scheduled to be completed.

   A. Strongly agree

   B. Agree

   C. Not sure

   D. Disagree

   E. Strongly disagree

Using the answer key that follows, please complete Exhibit 3.

▶ Answer Key

The assignment of the points is as follows:

Integrated Processes						
Question	Points					
1	A. 2	B. 2	C. 4	D. 2	E. 4	F. 5
2	A. 0	B. 0	C. 1	D. 3	E. 4	F. 5
3	A. 0	B. 0	C. 3	D. 4	E. 5	F. 5
4	A. 0	B. 1	C. 3	D. 4	E. 5	F. 5
5	A. 0	B. 2	C. 2	D. 2	E. 5	
6	A. 0	B. 2	C. 4	D. 5		
7	A. 0	B. 5	C. 4	D. 2	E. 0	

Culture					
Question	Points				
8	A. 0	B. 2	C. 3	D. 5	
9	A. 1	B. 3	C. 4	D. 4	E. 5	F. 5
10	A. 1	B. 5	C. 4	D. 0	
11	A. 3	B. 3	C. 3	D. 5	E. 0	F. 4
12	A. 1	B. 5	C. 5	D. 3	
13	A. 2	B. 3	C. 4	D. 5	E. 4
14	A. 2	B. 3	C. 4	D. 5	E. 5

Management Support					
Question	Points				
15	A. 1	B. 5	C. 5	D. 5	E. 0
16	A. 2	B. 3	C. 5	D. 0	E. 2
17	A. 3	B. 2	C. 5	D. 1	E. 0
18	A. 2	B. 3	C. 5	D. 0	E. 3
19	A. 4	B. 1	C. 2	D. 5	
20	A. 1	B. 1	C. 3	D. 4	E. 5
21	A. 1	B. 3	C. 5	D. 4	E. 4

Training and Education					
Question	Points				
22	A. 1	B. 3	C. 5	D. 5	E. 5
23	A. 0	B. 2	C. 4	D. 5	E. 5
24	A. 0	B. 3	C. 4	D. 5	E. 0
25	A. 2	B. 3	C. 4	D. 5	
26	A. 2	B. 1	C. 2	D. 3	E. 5
27	A. 0	B. 1	C. 3	D. 5	E. 5
28	A. 0	B. 1	C. 3	D. 4	E. 5

Informal Project Management					
Question	Points				
29	A. 2	B. 4	C. 5	D. 1	E. 0
30	A. 0	B. 3	C. 4	D. 5	E. 0
31	A. 5	B. 2	C. 3	D. 1	E. 0
32	A. 3	B. 5	C. 4	D. 2	E. 1
33	A. 2	B. 3	C. 4	D. 5	E. 0
34	A. 4	B. 5	C. 3	D. 1	E. 0
35	A. 3	B. 4	C. 3	D. 5	

Behavioral Excellence						
Question	Points					
36	A. 1	B. 2	C. 4	D. 5	E. 0	
37	A. 5	B. 1	C. 4	D. 2		
38	A. 3	B. 5	C. 5	D. 5	E. 4	
39	A. 2	B. 2	C. 2	D. 5	E. 3	
40	A. 3	B. 3	C. 5	D. 1		
41	A. 5	B. 3	C. 4	D. 1		
42	A. 5	B. 4	C. 2	D. 1	E. 1	

■ Exhibit 3

Determine your points for each of the questions, and complete the following:

A. Points for integrated processes (questions 1–7): _____

B. Points for culture (questions 8–14): _____

C. Points for management support (questions 15–21): _____

D. Points for training and education (questions 22–28): _____

E. Points for informal project management (questions 29–35): _____

F. Points for behavioral excellence (questions 36–42): _____

TOTAL: _____

► Explanation of Points for Level 3

Each of the six areas is a component of the Hexagon of Excellence discussed in Level 3. The total points can be interpreted as follows:

Points	Interpretation
169–210	Your company compares very well to the companies discussed in this text. You are on the right track for excellence, assuming that you have not achieved it yet. Continuous improvement will occur.
147–168	Your company is going in the right direction, but more work is still needed. Project management is not totally perceived as a profession. It is also possible that your organization simply does not fully understand project management. Emphasis is probably more toward being non–project-driven than project-driven.
80–146	The company is probably just providing lip service to project management. Support is minimal. The company believes that it is the right thing to do, but has not figured out the true benefits or what they, the executives, should be doing. The company is still a functional organization.
Below 80	Perhaps you should change jobs or seek another profession. The company has no understanding of project management, nor does it appear that the company wishes to change. Line managers want to maintain their existing power base and may feel threatened by project management.

Level 3 focuses heavily on the organization's culture and support for project management. Customization for Level 3 is relatively easy as long as the customized questions follow the categories in the Hexagon of Excellence. If customization follows the categories, the grading system may not need to be changed.

The amount of customization may be based on the size of the company and whether profit and loss are assigned to each project. If only five of the six categories are needed, then additional questions can be added to the other categories or a completely new category can be introduced.

This level may have to go through major changes if the assessments are being done and the firm is a heavy use of the agile or Scrum approach. Modifications for agile and Scrum will be discussed later in this book.

◀ **CHAPTER EIGHT** ▶

Level 4: Benchmarking

▶ Introduction

Project management benchmarking is the process of continuously comparing the project management practices of your organization with the practices of leaders anywhere in the world; its goal is to gain information to help you improve your own performance. The information obtained through benchmarking might be used to help you improve your processes and the way in which those processes are executed, or the information might be used to help your company become more competitive in the marketplace.

Benchmarking is a continuous effort of analysis and evaluation. Care must be taken in deciding what to benchmark. It is impossible and impractical to evaluate every aspect of project management. It is best to decide on those few critical success factors that must go right for your business to flourish. For project management benchmarking, the critical success factors are usually the key business processes and how they are integrated. If these key success factors do not exist, then the organization's efforts may be hindered.

Deciding what information to benchmark against is usually easier than obtaining that information. Locating some information will require a critical search. Some information may be hard to find. Some information you would find helpful might not be available for release because the organization that has it views it as proprietary. Identifying the target companies against which you should benchmark may not be as easy as you believe.

Benchmarking has become common since it was first popularized by Xerox during the 1980s. Benchmarking is an essential ingredient for those companies that have won the prestigious Malcolm Baldrige National Quality Award. Most of these award winners seem willing to readily share their project management experiences. Unfortunately, there are some truly excellent companies in project management that have not competed for these awards because they do not want their excellence displayed. They view it as a competitive advantage.

Benchmarking for project management can be accomplished through surveys, questionnaires, attending local chapter meetings of the Project Management Institute (PMI®), and attending conferences and symposiums. Personal contacts often provide the most valued sources of information.

There is a so-called *code of conduct* for benchmarking:

- Keep the benchmarking process legal.
- Do not violate rules of confidentiality.
- Sharing information is a two-way street.
- Be willing to sign a nondisclosure form.
- Do not share any information received with a third party without written permission.
- Emphasize guidelines and checklists, but avoid asking for forms that may be highly sensitive.

Benchmarking should not be performed unless your organization is willing to make changes. The changes must be part of a structured process that includes evaluation, applicability, and risk management. Benchmarking is part of the strategic planning process for project management that results in an action plan ready for implementation.

▶ Characteristics

Level 4 is the level where the organization realizes that its existing project management approach can be improved. The complexity rests in figuring out how to achieve that improvement. For project-driven companies, continuous improvement is a means to maintain or improve on a competitive advantage. Continuous improvement is best accomplished through continuous benchmarking. The company must decide whom to benchmark and what to benchmark.

Level 4 has certain characteristics, as shown in Figure 8.1:

- The organization must establish a project management office (PMO) or a center of excellence (COE) for project management. This is the focal position in the company for project management knowledge.
- The PMO or COE must be dedicated to the project management improvement process along with other activities. This is usually accomplished with full-time, dedicated personnel.
- Benchmarking must be made against both similar and nonsimilar industries. In today's world, a company with five years of experience in project management could easily surpass the capabilities of a company that has used project management for 20 years or more.
- The company should perform both quantitative and qualitative benchmarking. Quantitative benchmarking analyzes processes and methodologies, whereas qualitative benchmarking looks at project management applications.

Benchmarking
• Establishment of a project management office (PMO) or a center of excellence (COE)
• Dedication to benchmarking
• Looking at both similar and nonsimilar industries
• Quantitative benchmarking (processes and methodologies)
• Qualitative benchmarking (cultures)

Figure 8.1 Characteristics of Level 4.

▶ The Project Office or Center of Excellence

When companies reach Level 4, they are committed to project management across the entire organization. Project management knowledge is now considered as essential for the survival of the firm. To centralize the knowledge regarding project management, organizations have created a project management office (PMO) or a center of excellence (COE) for project management.

Major responsibilities for a PMO/COE include:

- A strategic planning focal point of project management

- An organization dedicated to benchmarking for project management

- An organization dedicated to continuous improvements in project management

- An organization that provides mentorship for inexperienced project managers

- A centralized data bank of lessons learned, possibly with a best practices library

- An organization for sharing project management ideas and experiences

- A hotline for problem-solving that does not automatically inform senior management

- An organization for creating project management standards and processes

- A focal point for centralized planning and scheduling activities

- A focal point for centralized cost control and reporting

- An organization to assist Human Resources in the creation of a project management career path

- An organization to assist Human Resources in developing a project management training curriculum

Most companies view the PMO and the COE as being two names for the same thing. There are, however, fundamental differences, as shown in Table 8.1. Despite the responsibilities, companies struggle with the organizational reporting location of the PMO/COE. There appears to be agreement that the location should be at the senior levels of management (see Figure 8.2).

Table 8.1 Project office versus center of excellence.

Project office	Center of excellence
Permanent line function for projects	May be a formal or informal manager committee (may be part time)
Focus on internal lessons-learned activities	Focuses on external benchmarking
Champion for the implementation of the methodology	Champion for continuous improvement and benchmarking
Expertise in the use of project management tools	Expertise in the identification of project management tools

Figure 8.2 Simplified PMO organizational chart.

▶ Benchmarking Opportunities

Historically, benchmarking has been accomplished using two approaches: competitive benchmarking and process benchmarking. Competitive benchmarking concentrates on deliverables and quantitative critical success factors. Process benchmarking focuses on process performance and functionality. Process benchmarking is most closely aligned to project management. For simplicity's sake, we will consider only process improvement benchmarking. We can break it down into quantitative (i.e., integration) process improvement opportunities and qualitative process improvement opportunities.

Figure 8.3 shows the quantitative process improvement opportunities, which center on enhancements due to integration opportunities. The five major areas identified in Figure 8.3 are the five integrated processes described in Level 3 of the project management maturity model (PMMM).

Figure 8.4 shows the qualitative process improvement opportunities, which center on applications and further changes to the corporate culture. Included in the qualitative process improvement activities are:

- *Corporate acceptance:* This includes getting the entire organization to accept a project management approach. Pockets of project management support tend to hinder rapid acceptance of project management by the rest of the organization. To obtain corporate acceptance, you must:

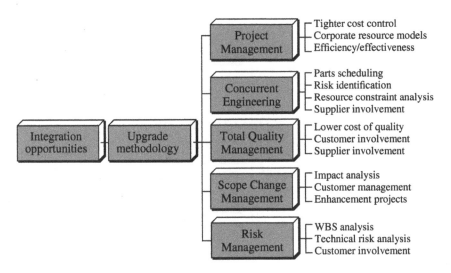

Figure 8.3 Quantitative process improvement opportunities (generic integrated process strategies).

- Increase the usage and support of existing users

- Attract new internal users, those who have been providing resistance to project management

- Discourage the unnecessary development of parallel methodologies, which can create further pockets of project management. This is done by showing the added costs of parallelization.

- Emphasize the present and future benefits to the corporation that will result from using a singular methodology as a starting point to project management maturity.

- *Integrated processes:* This is a recognition that the singular methodology can be enhanced further by integrating other existing processes into the singular methodology. Typically, this includes business processes such as capital budgeting, feasibility studies, cost-benefit analyses, and return-on-investment analyses. New processes that could be integrated include supply chain management.

- *Enhanced benchmarking:* Everyone tends to benchmark against the best within their own industry, but benchmarking against nonsimilar industries can be just as fruitful. An aerospace company spent over 10 years benchmarking only against other aerospace companies. During the mid-1990s, the firm began benchmarking against non-aerospace firms, and found that these firms had developed outstanding methodologies with capabilities exceeding those of the aerospace firm.

- *Software enhancements:* Although off-the-shelf software packages exist, most firms still need some type of customization. This can be done through internal upgrades for customization or by new purchases, with the software vendor developing the customization.

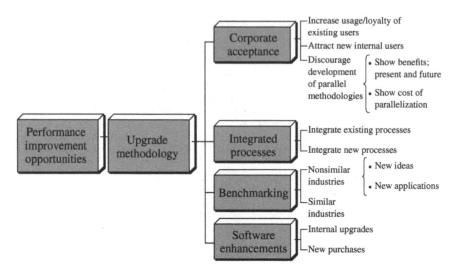

Figure 8.4 Qualitative process improvement opportunities (generic performance improvement strategies).

▶ Roadblocks

There also exist roadblocks to completing Level 4 and reaching Level 5, as shown in Figure 8.5. The singular methodology created in Level 3 was developed internally within the company. Benchmarking may indicate that improvements can be made. The original architects of the singular methodology may resist change with arguments such as, "It wasn't invented here," or "It does not apply to us." Another form of resistance is the argument that you have benchmarked against the wrong industry.

People are inherently fearful of change, and benchmarking opens the door for unexpected results to surface. Sooner or later, everyone realizes that benchmarking is a necessity for company survival. It is at this junction that a serious commitment to benchmarking occurs.

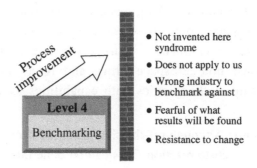

Figure 8.5 Roadblocks to completion of Level 4.

▶ Advancement Criteria

Four key actions are required by the organization to advance to Level 5, the final level. These actions are as follows:

- Create an organization dedicated to benchmarking.
- Develop a project management benchmarking process.
- Decide what to benchmark and against whom to benchmark.
- Recognize the benefits of benchmarking.

The successful completion of Level 4 is accompanied by a low degree of difficulty. Since the organization may have already accepted the idea of a singular methodology, it is a low risk to expect the employees to accept changes. They now know that change is inevitable.

▶ Assessment Instrument for Level 4

On the next several pages you will find 25 questions concerning how mature you believe your organization to be. Beside each question, circle the number that corresponds to your opinion. In the example below, your choice would have been Slightly Agree:

−3 Strongly Disagree

−2 Disagree

−1 Slightly Disagree

0 No Opinion

(+1) Slightly Agree

+2 Agree

+3 Strongly Agree

Example: (−3, −2, −1, 0, (+1), +2, +3)

The row of numbers from −3 to +3 will be used later for evaluating the results. After answering question 25, you will grade the exercise.

▶ Questions

The following 25 questions involve benchmarking. Please answer each question as honestly as possible. Circle the answer you feel is correct, not the answer you believe the question is seeking out.

1. Our benchmarking studies have found companies with tighter cost-control processes than we use.

(−3 −2 −1 0 +1 +2 +3)

2. Our benchmarking studies have found companies with better impact analysis during scope change control.

(−3   −2   −1   0   +1   +2   +3)

3. Our benchmarking studies have found that companies are performing risk management by analyzing the detailed level of the work breakdown structure (WBS).

(−3   −2   −1   0   +1   +2   +3)

4. Our benchmarking studies investigate supplier involvement in project management activities.

(−3   −2   −1   0   +1   +2   +3)

5. Our benchmarking studies investigate customer involvement in project management activities.

(−3   −2   −1   0   +1   +2   +3)

6. Our benchmarking studies investigate how to obtain increased loyalty/usage of our project management methodology.

(−3   −2   −1   0   +1   +2   +3)

7. Our benchmarking efforts look at industries in the same business area as our company.

(−3   −2   −1   0   +1   +2   +3)

8. Our benchmarking efforts look at nonsimilar industries (i.e., industries in different business areas).

(−3   −2   −1   0   +1   +2   +3)

9. Our benchmark efforts look at nonsimilar industries to seek out new ideas and new applications for project management.

(−3   −2   −1   0   +1   +2   +3)

10. Our benchmarking efforts look at other companies' concurrent engineering activities to see how they perform parts scheduling and tracking.

(−3   −2   −1   0   +1   +2   +3)

11. Our benchmarking efforts have found other companies that are performing resource-constraint analyses.

(−3   −2   −1   0   +1   +2   +3)

12. Our benchmarking efforts look at the way other companies manage their customers during the scope change management process.

(−3   −2   −1   0   +1   +2   +3)

13. Our benchmarking efforts look at the way other companies involve their customers during risk-management activities.

(−3   −2   −1   0   +1   +2   +3)

14. Our benchmarking efforts look at software enhancements through internal upgrades.

    (−3   −2   −1   0   +1   +2   +3)

15. Our benchmarking efforts look at software enhancements through new purchases.

    (−3   −2   −1   0   +1   +2   +3)

16. Our benchmarking efforts look at the way other companies attract new, internal users to their methodology for project management.

    (−3   −2   −1   0   +1   +2   +3)

17. Our benchmarking efforts focus on how other companies perform technical risk management.

    (−3   −2   −1   0   +1   +2   +3)

18. Our benchmarking efforts focus on how other companies obtain better efficiency and effectiveness of their project management methodology.

    (−3   −2   −1   0   +1   +2   +3)

19. Our benchmarking efforts focus on how to obtain a lower cost of quality.

    (−3   −2   −1   0   +1   +2   +3)

20. Our benchmarking efforts look at the way other companies perform risk management during concurrent engineering activities.

    (−3   −2   −1   0   +1   +2   +3)

21. Our benchmarking efforts look at the way other companies use enhancement projects as part of scope change management.

    (−3   −2   −1   0   +1   +2   +3)

22. Our benchmarking efforts look at ways of integrating existing processes into our singular methodology.

    (−3   −2   −1   0   +1   +2   +3)

23. Our benchmarking efforts look at ways other companies have integrated new methodologies and processes into their singular methodology.

    (−3   −2   −1   0   +1   +2   +3)

24. Our benchmarking efforts look at the way other companies handle or discourage the development of parallel methodologies.

    (−3   −2   −1   0   +1   +2   +3)

25. Our benchmarking efforts seek out other companies' use of corporate resource models.

    (−3   −2   −1   0   +1   +2   +3)

An answer sheet follows. Please complete Exhibit 4.

■ Exhibit 4

Each response you circled in questions 1–25 had a column value between −3 and +3. In the appropriate spaces below, place the circled value (between −3 and +3) beside each question.

Quantitative Benchmarking

1. _____
2. _____
3. _____
4. _____
5. _____
10. _____
11. _____
12. _____
13. _____
17. _____
18. _____
19. _____
20. _____
21. _____
25. _____

TOTAL _____

Qualitative Benchmarking

6. _____
7. _____
8. _____
9. _____
14. _____
15. _____
16. _____
22. _____
23. _____
24. _____

TOTAL _____

Quantitative benchmarking total: _____
Qualitative benchmarking total: _____
Combined total: _____

▶ Explanation of Points for Level 4

This exercise measures two items: Is your organization performing benchmarking? And, if so, are you emphasizing quantitative or qualitative benchmarking?

Quantitative benchmarking investigates improvements to the methodology and processes. Scores greater than 25 are excellent and imply that your organization is committed to quantitative benchmarking. Scores of 10 or less indicate a lack of commitment or that the organization does not understand how to benchmark or against whom to benchmark. Scores from 11 to 24 indicate that some benchmarking may be taking place, but a PMO or COE is not in place as yet.

Qualitative benchmarking looks more at applications benchmarking and how the culture executes the methodology. Scores greater than 12 are excellent. Scores of 5 or less indicate that not enough emphasis is placed on the "soft side" of benchmarking. Scores from 6 to 11 are marginally acceptable.

Combined scores (i.e., quantitative and qualitative) of 37 or more imply that your organization is performing benchmarking well. The right information is being considered, and the right companies are being targeted. The balance between quantitative and qualitative benchmarking is good. The company probably has a COE or PMO in place.

▶ Opportunities for Customizing Level 4

The questions in the benchmarking assessments can change if the firm is looking for specific information as part of benchmarking. The number of assessment questions can increase or decrease for quantitative and qualitative benchmarking, in which case the grading would have to be adjusted.

Other categories could also be included, such as industry benchmarking or benchmarking against specific companies such as those of a certain size. But the intent with this level and the assessment questions is to make sure the firm at least realizes the importance of benchmarking. As companies mature in project management, benchmarking may change from industry or competitive benchmarking to world-class benchmarking.

Level 4 can begin as early as Level 1. If this is done, then the information received can be valuable for Levels 2 and 3, and minimize downstream changes.

CHAPTER NINE

Level 5: Continuous Improvement

▶ Characteristics

At the previous level, the organization began benchmarking against other companies. At Level 5, the organization evaluates the information learned during benchmarking and implements the changes necessary to improve the project management process. It is at this level that the company comes to the realization that excellence in project management is a never-ending journey.

There are four characteristics of Level 5, as shown in Figure 9.1:

- The organization must create lessons-learned files from the debriefing sessions at the end of each project. Case studies on each project, discussing mistakes made and knowledge learned, are critical so that mistakes are not repeated.

- The knowledge learned on each project must be transferred to other projects and teams. This can be accomplished through quarterly or semiannual lessons-learned forums or from lessons-learned case studies discussed in training programs.

- The company must recognize that a mentorship program should be put in place to groom future project managers. Knowledge and lessons-learned information can be transmitted through the mentorship program as well. The mentorship program is best administered through a project management office (PMO) or a center of excellence (COE).

- The final characteristic of Level 5 is a corporate-wide understanding that strategic planning for project management is a continuous, ongoing process.

Documenting project results in lessons-learned files and the preparation of case studies can be difficult to implement. People learn from both successes and failures. One executive commented that the only true project failures are the ones from which we learned nothing. Another executive commented that project debriefings are a waste of time unless we learn something from them.

> ## Continuous improvement
>
> - Lessons-learned files
> - Knowledge transfer
> - COE/PMO mentorship program
> - Strategic planning for project management

Figure 9.1 Characteristics of Level 5.

Documenting successes is easy. Documenting mistakes is more troublesome because people do not want their names attached to mistakes for fear of retribution. Company employees still know which individuals worked on which projects, even when the case study is disguised. A strong corporate culture is needed to make documenting mistakes work effectively as a learning experience.

▶ Continuous Improvement Areas

Project management methodologies must undergo continuous improvement. This may be strategically important to stay ahead of the competition. Continuous improvements to a methodology can be internally driven by factors such as better software availability, a more cooperative corporate culture, or simply training and education in the use of project management practices. Externally driven factors include relationships with customers and suppliers, legal factors, social factors, technological factors, and even political factors.

Five areas for continuous improvement to the project management methodology are shown in Figure 9.2 and in the following:

Existing Process Improvements

- *Frequency of use:* Has prolonged use of the methodology made it apparent that changes can be made?

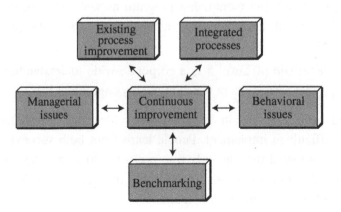

Figure 9.2 Factors to consider for continuous improvement.

- *Access to customers:* Can you improve the methodology to get closer to your customers?

- *Substitute products:* Are there new products (i.e., software) in the marketplace that can replace and improve part of your methodology?

- *Better working conditions:* Can changes in the working conditions cause you to eliminate parts of the methodology (i.e., paperwork requirements)?

- *Better use of software:* Will new or better use of the software allow you to eliminate some of your documentation and reports?

Integrated Process Improvements

- *Speed of integration:* Are there ways to change the methodology to increase the speed of integrating activities?

- *Training requirements:* Have changes in training requirements mandated changes in your methodology?

- *Corporate-wide acceptance:* Should the methodology change in order to obtain corporate-wide acceptance?

Behavioral Issues

- *Changes in organizational behavior:* Have changes in behavior mandated methodology changes?

- *Cultural changes:* Has your culture changed (i.e., to a cooperative culture) such that the methodology can be enhanced?

- *Management support:* Has management support improved to a point where fewer gate reviews are required?

- *Impact on informal project management:* Is there enough of a cooperative culture that informal project management can be used to execute the methodology?

- *Shifts in power and authority:* Do authority and power changes mandate a looser or a more rigid methodology?

- *Safety considerations:* Have safety or environmental changes occurred that will impact the methodology?

- *Overtime requirements:* Do new overtime requirements mandate an updating of forms, policies, or procedures?

Benchmarking

- *Creation of a project management COE:* Do you now have a core group responsible for benchmarking?

- *Cultural benchmarking:* Do other organizations have better cultures than you do in project management execution?

- *Process benchmarking:* What new processes are other companies integrating into their methodology?

Managerial Issues

- *Customer communications:* Have there been changes in the way you communicate with your customers?

- *Resource capability versus needs:* If your needs have changed, what has happened to the capabilities of your resources?

- *Restructuring requirements:* Has restructuring caused you to change your sign-off requirements?

- *Growing pains:* Does the methodology have to be updated to include your present growth in business (i.e., tighter or looser controls)?

These five factors provide a company with a good framework for continuous improvement. The benefits of continuous improvement include:

- Better competitive positioning
- Corporate unity
- Improved cost analysis
- Customer value added
- Better management of customer expectations
- Ease of implementation

▶ The Never-Ending Cycle

Given the fact that maturity in project management is a never-ending journey, we can define excellence in project management as a never-ending cycle of benchmarking–continuous improvement–singular methodology enhancement, as shown in Figure 9.3.

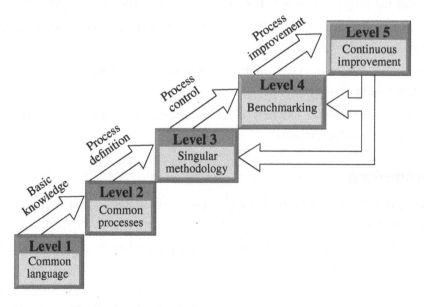

Figure 9.3 The five levels of maturity.

This implies that Levels 3, 4, and 5 of the PMMM are repeated over and over again. This also justifies our statement of the need for overlapping levels.

▶ Examples of Continuous Improvement

As more and more industries accept project management as a way of life, continuous improvement opportunities in project management practices have arisen at an astounding rate. What is even more important is the fact that companies are sharing their accomplishments with other companies during benchmarking activities.

Ten recent interest areas are included in this chapter:

- Developing effective procedural documentation
- Project management methodologies
- Continuous improvement
- Capacity planning
- Competency models
- Managing multiple projects
- End-of-phase review meetings
- Strategic selection of projects
- Portfolio selection of projects
- Horizontal accounting

These 10 topics appear to be the quickest to change. However, as project management evolves, other topics, such as those discussed in Chapter 11, may replace some of these.

▶ Developing Effective Procedural Documentation

Previously, we showed the necessity to develop processes and ultimately a singular methodology for project management. Project management methodologies require a project management information system (PMIS), which is based on procedural documentation. The procedural documentation can be in the form of policies, procedures, guidelines, forms and checklists, or a combination of these. Good procedural documentation will accelerate the project management maturity process, foster support at all levels of management, and greatly improve project communications. The type of procedural documentation selected can change over the years and is heavily biased toward whether you wish to manage more formally or informally. In any event, procedural documentation supports effective communications, which in turn provides for better interpersonal skills.

An important facet of any project management methodology is to provide the people in the organization with procedural documentation on how to conduct project-oriented activities and how to communicate in such a multidimensional environment.

The project management policies, procedures, forms, guidelines, templates, and checklists can provide some of these tools for delineating the process, as well as a format for collecting, processing, and communicating project-related data in an orderly, standardized format. Project planning and tracking, however, involve more than just the generation of paperwork. They require the participation of the entire project team, including support departments, subcontractors, and top management. This involvement of the entire team fosters a unifying team environment. This unity, in turn, helps the team focus on the project goals and, ultimately, fosters each team member's personal commitment to accomplishing the various tasks within time and budget constraints. The specific benefits of procedural documents, including forms and checklists, are that they help to:

- Provide guidelines and uniformity.
- Encourage useful, but minimum, documentation.
- Communicate clearly and effectively.
- Standardize data formats.
- Unify project teams.
- Provide a basis for analysis.
- Document agreements for future reference.
- Refuel commitments.
- Minimize paperwork.
- Minimize conflict and confusion.
- Delineate work packages.
- Bring new team members on board.
- Build an experience track and method for future projects.

Done properly, the process of project planning must involve both the performing and the customer organizations. This involvement creates new insight into the intricacies of a project and its management methods. It also leads to visibility of the project at various organizational levels, management involvement, and support. It is this involvement at all organizational levels that stimulates interest in the project and the desire for success, and fosters a pervasive reach for excellence that unifies the project team. It leads to commitment toward establishing and reaching the desired project objectives and to a self-forcing management system where people want to work toward these established objectives.

Not all procedural documentation will be used on every project, as companies embrace agile and Scrum. As such, project personnel will be given the freedom to decide which documentation is critical for a project.

■ The Challenges

Despite all of these benefits, management is often reluctant to implement or fully support a formal project management system. Management concerns often center on four issues: overhead burden, start-up delays, stifled creativity, and reduced self-forcing control. First, the introduction of more organizational formality via policies, procedures, and forms might cost some money, plus additional funding will be needed to support and maintain the system. Second, the system is seen, especially by action-oriented managers, as causing undesirable start-up delays by requiring the putting of certain stakes into the ground, in terms of project definition, feasibility, and organization, before the detailed implementation can start. Third and fourth, the system is often perceived as stifling creativity and shifting project control from the responsible individual to an impersonal process that enforces the execution of a predefined number of procedural steps and forms without paying attention to the complexities and dynamics of the individual project and its possibly changing objectives.

The comment of one project manager may be typical for many situations: "My support personnel feel that we spend too much time planning a project upfront; it creates a very rigid environment that stifles innovation. The only purpose seems to be establishing a basis for controls against outdated measures and for punishment rather than help in case of a contingency." This comment is echoed by many project managers. It's not a groundless attitude, for it also illustrates a potential misuse of formal project management systems: establishment of unrealistic controls and penalties for deviations from the program plan rather than help in finding solutions. Whether these fears are real or imaginary within a particular organization does not change the situation. It is the perceived coercion that leads to the rejection of the project management system. An additional concern is the lack of management involvement and funding to implement the project management system. Often the customer or sponsoring organization must also be involved and agree with the process for planning and controlling the project.

■ How to Make It Work

Few companies have introduced project management procedures with ease. Most have experienced problems ranging from skepticism to sabotage of the procedural system. Realistically, however, program managers do not have much of a choice, especially for larger, more complex programs. Every project manager who believes in project management has their own success story. It is interesting to note, however, that many have had to use incremental approaches to develop and implement their project management methodology.

Developing and implementing such a methodology incrementally is a multifaceted challenge to management. The problem is seldom one of understanding the techniques involved, such as budgeting and scheduling, but rather one of involving the project team in the process; getting their input, support, and commitment; and establishing a supportive environment. Furthermore, project personnel must have the feeling that the policies and procedures of the project management system facilitate communication, are flexible and adaptive to the changing environment, and provide an early warning system through which project personnel can obtain assistance rather than punishment in case of a contingency.

The procedural guidelines and forms of an established project management methodology can be especially useful during the project planning/definition phase. Not only do they help to delineate and communicate the four major sets of variables for organizing and managing the project—(1) tasks, (2) timing, (3) resources, and (4) responsibilities—but they also help to define measurable milestones, as well as report and review requirements. This, in turn, makes it possible to measure project status and performance and supplies the crucial inputs for controlling the project toward the desired results.

Developing an effective project management methodology takes more than just a set of policies and procedures. It requires the integration of these guidelines and standards into the culture and value system of the organization. Management must lead the overall efforts and foster an environment conducive to teamwork. The greater the team spirit, trust, commitment, and quality of information exchange among team members, the more likely it is that the team will develop effective decision-making processes, make individual and group commitments, focus on problem-solving, and operate in a self-forcing, self-correcting control mode. These are the characteristics that will support and pervade the formal project management system and make it work for you. When understood and accepted by the team members, such a system provides the formal standards, guidelines, and measures needed to direct a project toward specific results within the given time and resource constraints.

■ Established Practices

Although project managers may have the right to establish their own policies and procedures, many companies have taken the route of designing project-control forms that can be used uniformly on all projects to assist in the communications process. Project-control forms serve two vital purposes by establishing a common framework from which:

- The project manager will communicate with executives, functional managers, functional employees, and clients.
- Executives and the project manager can make meaningful decisions concerning the allocation of resources.

Success or failure of a project depends on the ability of key personnel to have sufficient data for decision-making. Project management is often considered to be both an art and a science. It is an art because of the strong need for interpersonal skills, and the project planning and control forms attempt to convert part of the "art" into a science.

Many companies tend not to realize until too late the necessity of good planning and control forms. Today, some of the larger companies with mature project management structures maintain a separate functional unit for forms control. This is quite common in aerospace and defense, but it is also becoming common practice in other industries. Yet some executives still believe that forms are needed only when the company grows to a point where a continuous stream of unique projects necessitates some sort of uniform control mechanism.

In some small or non–project-driven organizations, each project can have its own forms. But for most other organizations, uniformity is a must. Quite often, the actual design and selection of the forms is made by individuals other than the users. This can easily lead to disaster.

Large companies with a multitude of different projects do not have the luxury of controlling projects with three or four forms. There are different forms for planning, scheduling, controlling, authorizing work, and so on. It is not uncommon for companies to have 20 to 30 different forms, each dependent on the type of project, length of project, dollar value, type of customer reporting, and other such factors. Some companies have as many as 50 tools for project managers to use, and each tool has its own procedural documentation.

In project management, project managers are often afforded the luxury of being able to set up their own administration for the project, a fact that could lead to irrevocable long-term damage if each project manager were permitted to design his or her own forms for project control. Many times, this problem remains unchecked, and the number of forms grows exponentially with each project.

Executives can overcome this problem either by limiting the number of forms necessary for planning, scheduling, and controlling projects, or by establishing a separate department to develop the needed forms. Neither of these approaches is really practical or cost-effective. The best method appears to be the task force concept, where both managers and doers have the opportunity to interact and provide input. In the short run, this may appear to be ineffective and a waste of time and money. However, in the long run, there should be large benefits.

There is growth in the use of a flexible project management approach. This approach depends on the trust executives have in the project managers. With this approach, the organization establishes a library of forms and allows the project team to select which forms are appropriate for a project. The team may also design their own forms to meet a client's needs.

To be effective in traditional project management, the following ground rules can be used:

- Task forces should include managers as well as doers.
- Task-force members must be willing to accept criticism from other peers, superiors, and especially subordinates who must live with these forms.
- Upper-level management should maintain a rather passive (or monitoring) involvement.
- A minimum of signature approvals should be required for each form.
- Forms should be designed so that they can be updated periodically.
- Functional managers and project managers must be dedicated and committed to the use of the forms.

■ Categorizing the Broad Spectrum of Documents

The dynamic nature of project management and its multifunctional involvement create a need for a multitude of procedural documents to guide a project through the various phases and stages of integration. Especially for larger organizations, the challenge is not only to provide management guidelines for each project activity, but also to provide a coherent procedural framework within which project leaders from all disciplines can work and communicate with each other. Specifically, each policy or procedure must be consistent with and accommodating to the various other functions that interface with the project over its life cycle. This complexity of intricate relations is illustrated in Figure 9.4.

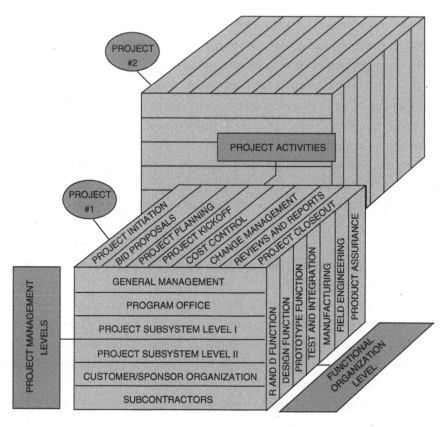

Figure 9.4 Interrelationship of project activities with various functional/organizational levels and project management levels.

Source: Reprinted from H. Kerzner and H. J. Thamhain, *Project Management Operating Guidelines* (New York: Van Nostrand Reinhold, 1985).

One simple and effective way of categorizing the broad spectrum of procedural documents is to utilize the work breakdown concept, as shown in Figure 9.5. This concept organizes the principal procedural categories along the lines of the principal project life-cycle phases. Each category is then subdivided into (1) general management guidelines, (2) policies, (3) procedures, (4) forms, and (5) checklists. If necessary, the concept can be extended an additional step to develop policies, procedures, forms, and checklists for the various project and functional sublevels of operation. Although this level of formality might be needed for very large programs, an effort should be made to minimize layering of policies and procedures, because the additional bureaucracy can cause new interface problems and additional overhead costs. For most projects, a single document covers all levels of project operations.

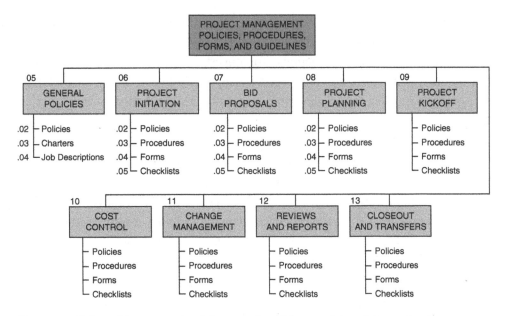

Figure 9.5 Categorizing procedural documents within a work breakdown structure.
Source: Reprinted from H. Kerzner and H. J. Thamhain, *Project Management Operating Guidelines* (New York: Van Nostrand Reinhold, 1985).

■ As We Mature …

As companies become more mature in executing the project management methodology, project management policies and procedures are discarded and replaced with guidelines, forms, and checklists. More flexibility is thus provided to the project manager. Unfortunately, reaching this stage takes time, because executives need to develop confidence in the ability of the project management methodology to work without the rigid controls provided by policies and procedures. All companies seem to go through the evolutionary stage of relying on policies and procedures before they advance to guidelines, templates, forms, and checklists.

▶ Project Management Methodologies

The ultimate purpose of any project management system is to drastically increase the likelihood that your organization will have a continuous stream of successfully managed projects. The best way to achieve this goal is with the development of a project management methodology. Good project management methodologies are based on guidelines and forms rather than policies and procedures. Methodologies must have enough flexibility that they can be adapted easily to each and every project. There are consulting companies that have created their own methodologies and that will try to convince you that the solution to most of your project management problems can be resolved with the purchase of their (often expensive) methodology. The primary goal of these consulting companies is turning problems into gold: your problems into their gold!

One major hurdle that any company must overcome when developing or purchasing a project management methodology is the fact that a methodology is nothing more than a sheet of paper with instructions. To convert this sheet of paper into a successful methodology, the company must accept, support, and execute the methodology. If this is going to happen, the methodology should be designed to support the corporate culture, not vice versa. It is a fatal mistake to purchase a canned methodology package that mandates you change your corporate culture to support it. If the methodology does not support the culture, the result will be a lack of acceptance of the methodology, sporadic use at best, inconsistent application of the methodology, poor morale, and perhaps even diminishing support for project management. What converts any methodology into a world-class methodology is its adaptability to the corporate culture.

There is no reason why organizations cannot develop their own methodologies. The amount of time and effort needed to develop a methodology will vary from company to company, based on such factors as the size and nature of the projects, the number of functional boundaries to be crossed, whether the organization is project-driven or non–project-driven, and competitive pressures.

▶ Continuous Improvement

All too often, complacency directs the decision-making process. This is particularly true of organizations that have reached some degree of excellence in project management and become self-satisfied. They often realize only too late that they have lost their competitive advantage. This occurs when organizations fail to recognize the importance of continuous improvement.

Figure 9.6 illustrates activities in a continuous improvement cycle. As companies begin to mature in project management and reach some degree of excellence, they achieve a sustained competitive advantage. Achieving this edge might very well be the single most important strategic objective of the firm. Once the firm has this sustained competitive advantage, it will then begin to exploit it.

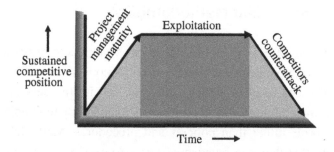

Figure 9.6 Activities in a continuous-improvement cycle.

Unfortunately, the competition will not sit by idly, watching you exploit your sustained competitive advantage; they will begin to counterattack. When they do, you may lose a large portion, if not all, of your sustained competitive advantage. To remain effective and competitive, your organization must recognize the need for continuous improvement, as shown in Figure 9.7. Continuous improvement allows a firm to maintain its competitive advantage even when its competitors counterattack.

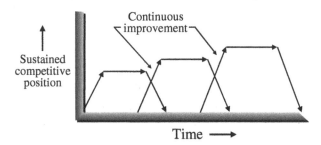

Figure 9.7 The need for continuous improvement.

▶ Capacity Planning

As companies become excellent in project management, the benefits of performing more work in less time and with fewer resources become readily apparent. The question, of course, is how much more work the organization can take on. Companies are now struggling to develop capacity-planning models to see how much new work can be undertaken within the existing human and nonhuman constraints.

Figure 9.8 illustrates the classical way that companies perform capacity planning. The approach shown holds true for both project- and non–project-driven organizations. The "planning horizon" line indicates the point in time for capacity planning. The "proposals" line indicates the resources needed for approved internal projects or a percentage (perhaps as much as 100 percent) for all work expected through competitive bidding. The combination of this line and the "manpower requirements" line, when compared against current staffing, provides an indication of capacity. This planning technique can be effective if performed early enough that training time is allowed for future manpower shortages.

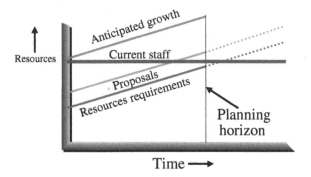

Figure 9.8 Classical capacity planning over time.

There is an important limitation to this process for capacity planning, however: Only human resources are considered. A more realistic method would be to use the strategy shown in Figure 9.9, which can also be applied to both project-driven and non–project-driven organizations. Using the approach in Figure 9.9, projects are selected based on such factors as strategic fit, profitability, consideration of who the customer is, and corporate benefits. The objectives for the projects selected are then defined in both business and technical terms, because there can be both business and technical capacity constraints.

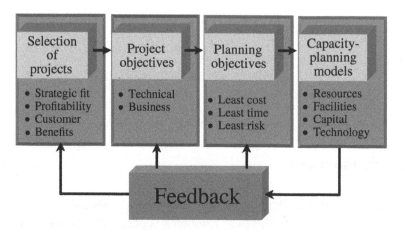

Figure 9.9 Capacity-planning activities.

The next step points up one critical difference between average companies and excellent companies. An excellent company will identify capacity constraints from the summation of the schedules and plans. Project managers will meet with project sponsors to determine the objective of the plan, which is different than the objective of the project. Is the objective of the plan to achieve the project's objective with the least cost, least time, or least risk? Typically, only one of these applies, whereas immature organizations believe that all three can be achieved on every project. This, of course, is unrealistic.

The final box in Figure 9.9 is now the determination of the capacity limitations. Previously, we considered only human resource capacity constraints. Now we realize that the critical path of a project can be constrained not only by available manpower but also by time, facilities, cash flow, and even technology. It is possible to have multiple critical paths on a project other than those identified by classical capacity planning. Each of these critical paths provides a different dimension to the capacity-planning models, and each of these constraints can lead to a different capacity limitation. As an example, manpower might limit you to taking on only four additional projects. Based on available facilities, however, you might only be able to undertake two more projects; and based on available technology, you might be able to undertake only one new project.

▶ Competency Models

Over the past decade, companies began replacing job descriptions with competency models. Job descriptions for project management tend to emphasize the deliverables and expectations from the project manager, whereas competency models emphasize the specific skills needed to achieve the deliverables.

Figure 9.10 shows the competency model for Eli Lilly. Project managers are expected to have competencies in three broad areas:[1]

- Scientific/technical skills
- Leadership skills
- Process skills

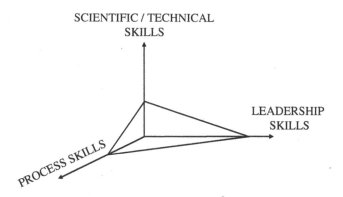

Figure 9.10 Competency model for Eli Lilly.

For each of the three broad areas, there are subdivisions or grade levels. A primary advantage of a competency model is that it allows the training department to develop customized project management training programs to satisfy the skill requirements. Without competency models, most training programs are generic rather than customized. Also, competency models make it easier for organizations to develop a complete training curriculum, rather than a single course.

Competency models focus on specialized skills in order to assist project managers in more efficiently utilizing their time. Figure 9.11, although theoretical, shows how, with specialized competency training, project managers might be able to increase their time effectiveness by reducing time robbers and rework. Unfortunately, time robbers and rework cannot always be eliminated, but they can be reduced.

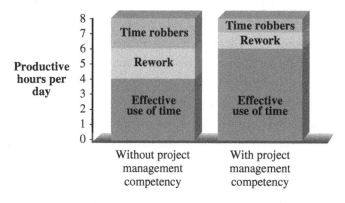

Figure 9.11 Core competency analysis.

---

[1] A detailed description of the Eli Lilly competency model can be found in H. Kerzner, *Project Management Best Practices: Achieving Global Excellence* (Hoboken, NJ: Wiley, 2018), 373–384.

As stated previously, competency models make it easier for companies to develop project management curricula, rather than simply single courses. This is shown in Figure 9.12. As companies mature in project management and develop a company-wide core competency model, an internal, custom-designed curriculum will be developed. Companies, especially large ones, will find it necessary to maintain a course architecture specialist on their staffs.

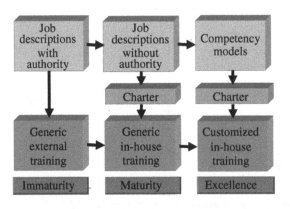

Figure 9.12 Competency models and training.

▶ Managing Multiple Projects

As organizations begin to mature in project management, there is a tendency toward wanting to manage multiple projects. This might entail either the company sponsoring the various projects, or each project manager managing multiple projects. Several factors support the managing of multiple projects.

First, the cost of maintaining a full-time project manager on all projects may be prohibitive. The magnitude and risks of the project dictate whether a full-time or part-time assignment is necessary. Assigning a project manager full-time on an activity that does not require it is an overmanagement cost. Overmanagement of projects was considered an acceptable practice in the early days of project management because we had little knowledge regarding how to handle risk management. Today, methods for risk management exist.

Second, line managers now share accountability with project managers for the successful completion of projects. Project managers are managing at the template levels of the work breakdown structure (WBS), with line managers accepting accountability for the work packages at the detailed WBS levels. Project managers now spend more of their time integrating work rather than planning and scheduling functional activities. With the line managers accepting more accountability, time may be available for project managers to manage multiple projects.

Third, senior management has come to the realization that they must provide high-quality training for project managers if they are to reap the benefits of managing multiple projects. Senior managers must also change the way they function as sponsors. There are six major areas where the corporation as a whole may have to change in order for the managing of multiple projects to succeed:

- *Prioritization:* If a project-prioritization system is in effect, it must be used correctly such that employee credibility in the system is realized. There are downside risks to a prioritization system. The project manager, having multiple projects to manage, may favor those projects having the highest priorities. It is possible that no prioritization system at all may be the best solution. Also, not every project needs to be prioritized. Prioritization can be a time-consuming effort.

- *Scope changes:* Managing multiple projects is almost impossible if sponsors/customers are allowed to make continuous scope changes. When managing multiple projects, the project manager must understand that the majority of the scope changes desired may have to be performed through enhancement projects rather than through a continuous scope-change effort on the original projects. A major scope change on one project could limit the project manager's available time to service other projects. Also, continuous scope changes will almost always be accompanied by reprioritization of projects, a further detriment to the management of multiple projects.

- *Capacity planning:* Organizations that support the management of multiple projects generally have tight control over resource scheduling. As a precondition, these organizations must have knowledge of capacity planning, theory of constraints, resource leveling, and resource-limited planning.

- *Project methodology:* Methodologies for project management range from rigid policies and procedures to more informal guidelines and checklists. When managing multiple projects, the project manager must be granted some degree of freedom. This necessitates guidelines, checklists, and forms. Formal project management practices create excessive paperwork requirements, thus minimizing the opportunities to manage multiple projects. The project size is also critical.

- *Project initiation:* Managing multiple projects has been going on for almost 50 years. One thing we have learned is that it can work well as long as the projects are in relatively different life-cycle phases. The demands on the project manager's time are different during each life-cycle phase. Therefore, for the project manager to effectively balance his or her time among multiple projects, it would be best for the sponsor not to have the projects begin at exactly the same time.

- *Organizational structures:* If the project manager is to manage multiple projects, then it is highly unlikely that the project manager will be a technical expert in all areas of all projects. Assuming that the accountability is shared with the line managers, the organization will most likely adopt a weak-matrix structure.

▶ End-of-Phase Review Meetings

For more than 30 years, end-of-phase review meetings were simply an opportunity for executives to rubber-stamp the project to continue. The meetings were used to give the executives some degree of comfort concerning project status. Only good news was presented by the project team.

Executives, from a selfish point of view, very rarely canceled projects. The executive was better off allowing the new product to be developed, even though the executive knew full well that the product would have no buyers or would be overpriced. Once the product was developed, the executive sponsor was off the hook. The onus now rested on the shoulders of the marketing group to find potential customers. If customers could not be found, obviously the problem was with marketing.

Today, end-of-phase review meetings take on a different dimension. First and foremost, executives are no longer afraid to cancel projects, especially if the objectives have changed, the objectives are unreachable, or the resources could be used on other activities that have a greater likelihood of success. Executives now spend more time assessing the risks in the future rather than focusing on accomplishments in the past.

Since project managers are now becoming more business-oriented, rather than technically oriented, they are expected to present information on business risks, reassessment of the benefit-to-cost ratio, and any business decisions that could affect the ultimate objectives. Simply stated, end-of-phase review meetings now focus more on business decisions than on technical decisions.

▶ Strategic Selection of Projects

What a company *wants* to do is not always what it *can* do. The critical constraint is normally the availability and quality of the critical resources. Companies usually have an abundance of projects they would like to work on; but, because of resource limitations, they have to develop a prioritization system for the selection of projects.

One commonly used selection process is the portfolio classification matrix shown in Figure 9.13. Each potential project undergoes a situational assessment for strengths, weaknesses, opportunities, and threats. The project is then ranked on the nine-square grid, based on its potential benefits and the quality of resources needed to achieve those benefits. The characteristics of the benefits appear in Figure 9.14, and the characteristics of the resources needed are shown in Figure 9.15.

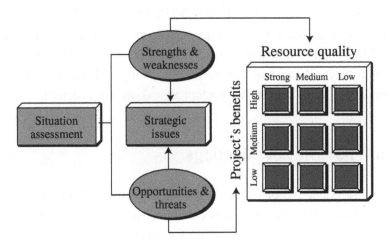

Figure 9.13 Portfolio classification matrix.

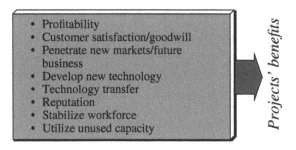

Figure 9.14 Potential benefits of a project.

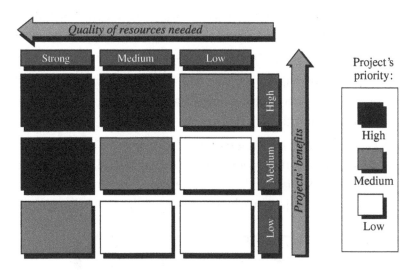

Figure 9.15 Characteristics of the resources needed to achieve a project's benefits.

Figure 9.16 Strategic importance of projects.

This classification technique allows for proper selection of projects, as well as providing the organization with the foundation for a capacity-planning model to see how much work the organization can take on. Companies usually have little trouble figuring out where to assign highly talented people. The model, however, provides guidance on how to make the most effective use of average and below-average individuals as well.

The boxes in the nine-square grid in Figure 9.13 can then be prioritized according to strategic importance, as shown in Figure 9.16. If resources are limited but funding is adequate, the boxes identified as high priority will be addressed first.

The nine-square grid in Figure 9.16 can also be used to identify the quality of the project management skills needed, in addition to the quality of functional employees. This is shown in Figure 9.17. As an example, the project managers with the best overall skills will be assigned to those projects that are needed to protect the firm's current position. Each of the nine cells in Figure 9.17 can be described as follows:

- *Protect position (high benefits and high quality of resources):* These projects may be regarded as vital to the survival of the firm. They mandate professional project management, possibly certified project managers, and that the organization consider project management as a career path position. Continuous improvement in project management is essential to make sure the methodology is the best it can be.

- *Protect position (high benefits and medium quality of resources):* Projects in this category may require a full-time project manager, but not necessarily a certified one. An enhanced project management methodology is needed, with emphasis on reinforcing vulnerable areas of project management.

- *Protect position (medium benefits and high quality of resources):* Emphasis in these projects is on training project managers, with special attention to their leadership skills. The types of projects here are usually efforts to add customer value rather than to develop new products.

- *Line management project management (high benefits and low quality of resources):* These projects are usually process-improvement efforts to support repetitive production. Minimum integration across functional lines is necessary, which allows line managers to function as project managers. These projects are characterized by short time frames.

- *Build selectively (medium benefits and medium quality of resources):* These projects are specialized, perhaps repetitive, and focus on a specific area of the business. Limited project management strengths are needed. Risk management may be needed, especially technical risk management.

- *Team leaders (low benefits but high quality of resources):* These are normally small, short-term R&D projects that require strong technical skills. Since minimal integration is required, scientists and technical experts will function as team leaders. Minimal knowledge of project management is needed.

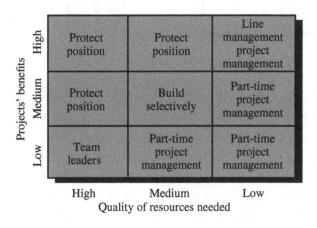

Figure 9.17 Strategic guide to allocating project resources.

- *Part-time project management (medium benefits and low quality of resources):* These are small capital projects that require only an introductory knowledge of project management. One project manager could end up managing multiple small projects.

- *Part-time project management (low benefits and medium quality of resources):* These are internal projects or very small capital projects. These projects have small budgets and perhaps a low to moderate risk.

- *Part-time project management (low benefits and low quality of resources):* These projects are usually planned by line managers but executed by project coordinators or project expediters.

▶ Portfolio Selection of Projects

Companies that are project-driven organizations must be careful about the type and quantity of projects they work on, due to the constraints on available resources. Because timing is often critical, it is not always possible to hire new employees and have them trained quickly enough, or to hire subcontractors, whose skills may well be questionable anyway.

Figure 9.18 shows a typical project portfolio.[2] Each circle represents a project. The location of each circle represents the quality of resources needed and the life-cycle phase the project is in. The size of the circle represents the magnitude of the achievable

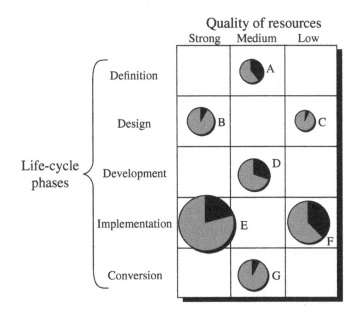

Figure 9.18 Basic portfolio.

---

[2] This type of portfolio was adapted from the life-cycle portfolio model used for strategic planning activities.

benefits, relative to those of other projects, and the "pie wedge" represents the percentage of the project completed thus far.

In Figure 9.18, Project A has relatively low benefits and uses medium quality of resources. Project A is in the definition phase. However, when Project A moves into the design phase, the quality of resources may change to low or high quality. Therefore, this type of chart has to be updated frequently.

Figures 9.19, 9.20, and 9.21 show three different types of portfolios. Figure 9.19 represents a high-risk project portfolio where high-quality resources are required on each project. This may be representative of a project-driven organization that has been awarded several highly profitable, large projects. This could also be a company that competes in the computer field, an industry that has short product life cycles and where product obsolescence occurs only six months downstream.

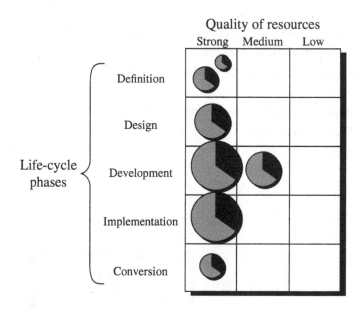

Figure 9.19 Typical high-risk project portfolio.

Figure 9.20 represents a conservative, profit-oriented project portfolio—say, that of an organization that works mainly on low-risk projects that require low-quality resources. This could be representative of project portfolio selection in a service organization, or even a manufacturing firm that has projects designed mostly for product enhancement.

Figure 9.21 shows a balanced portfolio with projects in each life-cycle phase and where all qualities of resources are being utilized, usually quite effectively. A very delicate juggling act is required to maintain this balance.

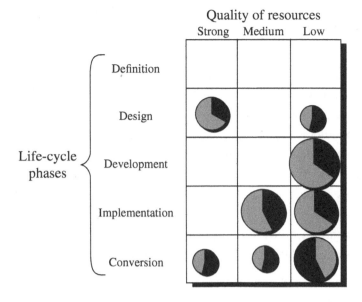

Figure 9.20 Typical conservative, profit-oriented project portfolio.

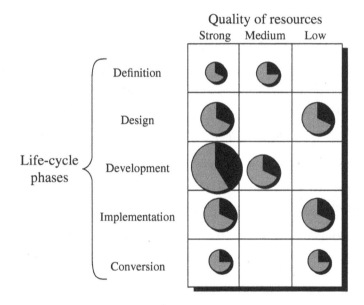

Figure 9.21 Typical balanced project portfolio.

▶ Horizontal or Project Accounting

In the early days of project management, project management was synonymous with scheduling. Project planning meant simply laying out a schedule, with very little regard for costs. After all, we know that costs will change (i.e., most likely increase) over the life of the project and that the final cost will never resemble the original budget. Therefore, why worry about cost control?

Recessions and poor economic times have put pressure on the average company to achieve better cost control. Historically, costs were measured on a vertical basis only. This created a problem in that project managers had no knowledge of how many hours were actually being expended in the functional areas to perform the assigned project activities. Standards were very rarely updated, and, if they were, it was usually without the project manager's knowledge.

Today, methodologies for project management mandate horizontal accounting using earned-value-measurement techniques. This is extremely important, especially if the project manager has the responsibility for profit and loss. Projects are now controlled through a series of charge numbers or cost-account codes assigned to all the work packages in the WBS.

Strategic planning for cost control on projects is a three-phase effort, as shown in Figures 9.22 through 9.24. The three phases are as follows:

- Phase I—Budget-based planning (Figure 9.22): This is the development of a project's baseline budget and cash flow based on reasonably accurate historical data. The historical databases are updated at the end of each project.

- Phase II—Cost/performance determination (Figure 9.23): Here, costs are determined for each work package, and actual costs are compared against actual performance in order to determine the true project status.

- Phase III—Updating and reporting (Figure 9.24): This involves the preparation of the necessary reports for the project team members, line managers, sponsors, and customer. At a minimum, the reports should address these questions:

  - Where are we today (time and cost)?

  - Where will we end up (time and cost)?

  - What problems and/or risks do we have now and will we have in the future, and what mitigation strategies have we come up with?

Good methodologies provide the framework for gathering the information to answer these questions.

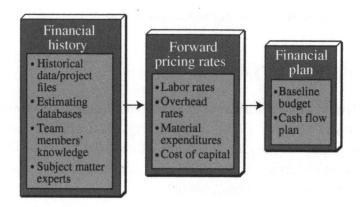

Figure 9.22 The evolution of integrated cost-schedule management. Phase I—Budget-based planning.

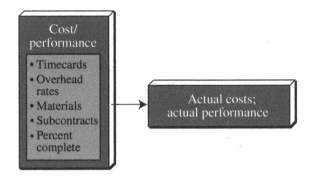

Figure 9.23 The evolution of integrated cost-schedule management. Phase II—Cost/performance determination.

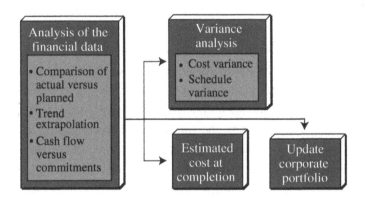

Figure 9.24 The evolution of integrated cost-schedule management. Phase III—Updating and reporting.

► Organizational Restructuring

Effective project management cultures are based on trust, communication, cooperation, and teamwork. When the basis of project management is strong, organizational structure becomes almost irrelevant. Restructuring an organization only to add project management is unnecessary and perhaps even dangerous. Companies may need to be restructured for other reasons, such as making the customer more important. But successful project management can exist within any structure, no matter how awful the structure looks on paper, as long as the culture of the company promotes teamwork, cooperation, trust, and effective communication.

The organizations of companies excellent in project management can take almost any form. Today, small- to medium-size companies sometimes restructure to pool management resources. Large companies tend to focus on the strategic business unit as the foundation of their structures. Many companies still follow matrix management. Any structure can work with project management as long as it has the following traits:

- The company is organized around nondedicated project teams.
- It has a flat organizational hierarchy.

- It practices informal project management.
- It does not consider the reporting level of project managers to be important.

The first point listed here may be somewhat controversial. Dedicated project teams have been a fact of life since the late 1980s. Although there have been many positive results from dedicated teams, there has also been a tremendous waste of manpower coupled with duplication of equipment, facilities, and technologies. Today, most experienced organizations believe that they are scheduling resources effectively so that multiple projects can make use of scarce resources at the same time. And, they believe, nondedicated project teams can be just as creative as dedicated teams, perhaps at a lower cost.

Although tall organizational structures with multiple layers of management were the rule when project management came on the scene in the early 1960s, today's organizations tend to be lean and mean, with fewer layers of management than ever. The span of control has been widened, and the result of that change has been mass confusion in some companies but complete success in others. The simple fact is that flat organizations work better. They are characterized by better internal communication, greater cooperation among employees and managers, and atmospheres of trust.

In addition, today's project management organizations, with only a few exceptions (purely project-driven companies), prefer to use informal project management. With formal project management systems, the authority and power of project managers must be documented in writing. Formal project management policies and procedures are required. And documentation is required for the simplest tasks. By contrast, in informal systems, paperwork is minimized. In the future, I believe that even totally project-driven organizations will develop more informal systems.

The reporting level for project managers has fluctuated between top-level and lower-level managers. As a result, some line managers have felt alienated over authority and power disagreements with project managers. In the most successful organizations, the reporting level has stabilized, and project managers and line managers today report at about the same level. Project management simply works better when the managers involved view each other as peers. In large projects, however, project managers may report higher up, sometimes to the executive level. For such projects, a project office is usually set up for project team members at the same level as the line managers with whom they interact daily.

To sum it all up, effective cross-functional communication, cooperation, and trust are bound to generate organizational stability. Let's hope that organizational restructuring on the scale we've seen in recent years will no longer be necessary.

▶ Career Planning

In organizations that successfully manage their projects, project managers are considered professionals and have distinct job descriptions. Employees traditionally are allowed to climb one of two career ladders: the management ladder or the technical ladder. (They cannot, however, jump back and forth between the two.) This presents a problem to project managers, whose responsibilities bridge the two ladders. To solve this problem, some organizations have created a third ladder, one that fills the gap between technology and management. It is a project management ladder, with the same opportunities for advancement as the other two. In some cases, this ladder becomes extremely important if the firm considers project management to be a strategic competency.

▶ Assessment Instrument for Level 5

The following 16 questions concern how mature you believe your organization to be with regard to Level 5. Beside each question, circle the number that corresponds to your opinion. In the example below, your choice would have been Slightly Agree.:

−3 Strongly Disagree

−2 Disagree

−1 Slightly Disagree

  0 No Opinion

(+1) Slightly Agree

+2 Agree

+3 Strongly Agree

Example: (−3   −2   −1   0   (+1)   +2   +3)

The row of numbers from −3 to +3 will be used later for evaluating the results. After answering question 16, you will grade the exercise by completing Exhibit 5.

▶ Questions

Answer the following questions based on continuous improvement changes over the past 12 months only. Circle the answer you feel is correct.

1. The improvements to our methodology have pushed us closer to our customers.

(−3   −2   −1   0   +1   +2   +3)

2. We have made software enhancements to our methodology.

(−3   −2   −1   0   +1   +2   +3)

3. We have made improvements that allowed us to speed up the integration of activities.

(−3   −2   −1   0   +1   +2   +3)

4. We have purchased software that allowed us to eliminate some of our reports and documentation.

(−3   −2   −1   0   +1   +2   +3)

5. Changes in our training requirements have resulted in changes to our methodology.

(−3   −2   −1   0   +1   +2   +3)

6. Changes in our working conditions (i.e., facilities, environment) have allowed us to streamline our methodology (i.e., paperwork reduction).

(−3   −2   −1   0   +1   +2   +3)

7. We have made changes to the methodology in order to get corporate-wide acceptance.

(−3   −2   −1   0   +1   +2   +3)

8. Changes in organizational behavior have resulted in changes to our methodology.

(−3   −2   −1   0   +1   +2   +3)

9. Management support has improved to the point where we now need fewer gates and checkpoints in our methodology.

(−3   −2   −1   0   +1   +2   +3)

10. Our culture is a cooperative culture to the point where informal rather than formal project management can be used, and changes have been made to the informal project management system.

(−3   −2   −1   0   +1   +2   +3)

11. Changes in power and authority have resulted in looser methodology (i.e., guidelines rather than policies and procedures).

(−3   −2   −1   0   +1   +2   +3)

12. Overtime requirements mandated change in our forms and procedures.

(−3   −2   −1   0   +1   +2   +3)

13. We have changed the way we communicate with our customers.

(−3   −2   −1   0   +1   +2   +3)

14. Because our projects' needs have changed, so have the capabilities of our resources.

(−3   −2   −1   0   +1   +2   +3)

15. (If your organization has restructured) Our restructuring caused changes in sign-off requirements in the methodology.

(−3   −2   −1   0   +1   +2   +3)

16. Growth of the company's business base has caused enhancements to our methodology.

(−3   −2   −1   0   +1   +2   +3)

■ Exhibit 5

Each response you circled in questions 1–16 had a column value between −3 and +3. In the appropriate spaces below, place the circled value (between −3 and +3) beside each question.

1. _____
2. _____
3. _____
4. _____
5. _____
6. _____
7. _____
8. _____
9. _____
10. _____
11. _____
12. _____
13. _____
14. _____
15. _____
16. _____
TOTAL: _____

The grading system for this exercise follows.

▶ Explanation of Points for Level 5

Scores of 20 or more are indicative of an organization committed to benchmarking and continuous improvement. These companies are probably leaders in their field. These companies will always possess more project management knowledge than both their customers and their competitors.

Scores from 10–19 are indicative that some forms of continuous improvement are taking place, but the changes may be occurring slowly. There may be resistance to some of the changes, most likely because of shifts in the power and authority spectrum.

Scores less than 10 imply a strong resistance to change or simply a lack of senior management support for continuous improvement. This most likely occurs in low technology, non–project-driven organizations where projects do not necessarily have a well-defined profit–loss statement. These organizations will eventually change only after pressure by their customers or an erosion of their business base.

▶ Opportunities for Customizing Level 5

The assessment questions in Level 5 are generic, such as improvements to the methodology. If the company has a list of improvements it wishes to make now or in the future, the questions can change.

The questions can also be customized for continuous improvement opportunities when dealing with clients and stakeholders. Customization can also include behavioral changes needed for improvements to the culture.

Chapter 11 discusses many of the changes taking place in project management as we begin using the concepts of PM 2.0 and PM 3.0. These topics can also be tested in the assessments.

CHAPTER TEN

Sustainable Competitive Advantage

▶ Introduction

To spend time and money developing a project management methodology or approach because you believe it is the right thing to do is a wasted effort. The better approach is to develop a methodology or approach with the intent of converting it into a sustainable competitive advantage. A sustainable competitive advantage not only placates your customers, it also puts pressure on your competitors to spend money to compete with you.

Sustainable competitive advantages can be determined for individual functional areas rather than for the entire company. As an example, consider Figure 10.1, which illustrates the efforts needed to achieve a sustained competitive advantage in research and development (R&D). As a company advances through the various stages of innovation, the technical risks will increase. The organization must have developed a good approach to the problem of assessing technical risks and must be willing to admit when a project should be cancelled because the resources could be allocated more effectively on other projects. Maintaining a competitive advantage requires a continuous stream of new and/or enhanced products or services. Risk management is an essential ingredient in the evaluation process.

As technical risks increase, so does the amount of money expended, as well as the requirement for superior technical ability. The technical skills required increase as you go from basic to applied research and on through development. Although some people may argue about the need for this increase in skill levels, the fact remains that a product that can be developed on a small laboratory bench may never be able to be mass-produced; or, even if it can be mass-produced, the quality may have to be degraded. Also, it is in development that you finally obtain the hard numbers as to whether the product can be manufactured at a competitive price.

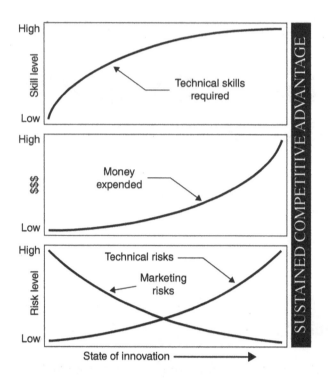

Figure 10.1 R&D efforts for a sustained competitive advantage.

Source: Reprinted from P. Rea and H. Kerzner, *Strategic Planning* (New York: Wiley, 1997), 105.

▶ Strategic Thrusts

As shown in Figure 10.2, there are four *strategic thrusts* that must be considered before your project management methodology can be turned into a sustainable competitive advantage. These strategic thrusts must be identified while the methodology is being designed and developed, not later. Developing a methodology and then having to make major changes to it because the strategic thrusts were not considered can waste time and money, and can also lower morale. Poor morale can cause the workers to lose faith in the methodology.

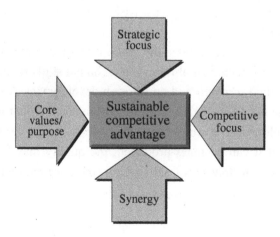

Figure 10.2 Strategic thrusts.

The first strategic thrust is the core values/purpose. The core values/purpose thrust describes the heart of the company, as well as the basic reason for its existence:

- *Core values:* A company usually has three to five core values: the timeless, passionately held guiding principles of the organization. At Procter & Gamble, for example, the core values are delivering consumer value, developing breakthrough innovation, and building strong brands. The core values for the Walt Disney Company might be imagination and wholesomeness, while at Nordstrom they could be service to the customer, trust, and products with style. Core values come from within the organization; they represent what the organization is at its very essence, as opposed to what it does from day to day.

- *Core purpose:* An organization's core purpose should last for at least 100 years; it is the organization's reason for being that goes beyond current products and services. For 3M, the core purpose is "to solve unsolved problems innovatively." For Hewlett-Packard, it is "to make technical contributions for the advancement and welfare of humanity." For McKinsey & Company, it is "to help leading corporations and governments to be more successful." For Merck, it is "to preserve and improve human life." And for the Walt Disney Company, it is "to make people happy." One approach to finding a core purpose is to ask five whys. Start with a description of the business, and ask, "Why is that important?" five times; after a few whys, you get to the very essence of the business.[1]

Generally speaking, all projects undertaken using the project management methodology must support the company's core values/purpose, which could very well be regarded as the most important strategic thrust.

The second strategic thrust in Figure 10.2 is the strategic focus. The strategic focus identifies the product/market element in which the organization competes. Three primary questions must be addressed in the strategic focus:

- Where will the organization compete? (What products are offered, and which markets are served, by segment or geographically?)

- Against whom will the organization compete? (Who is the competition?)

- How will the organization compete? (By product, by proper positioning, by functional strategy as channels of distribution, etc.?)

The answers to these three questions provide guidance regarding the quality and competencies of the resources and assets needed. Project management methodologies must be designed around the competencies of the resources.

The third strategic thrust is the competitive focus. Although this thrust has some similarities to the strategic focus thrust, there are other overriding factors. The

---

[1] D. A. Aaker, *Strategic Market Management*, 5th ed. (New York: Wiley, 1998), 28.

competitive focus emphasizes the differences between your organization and your major competitors. The differences can exist in such areas as:

- Product features
- Product design
- Product performance
- Product quality
- Products offered
- Value-added opportunities
- Brand name and image
- Cost-reduction opportunities (i.e., experience curves, labor rates)
- Strategic alliances and partnerships

These strategic competitive differences can give your methodology one step up on the competition.

The final strategic thrust in Figure 10.2 is synergy. Synergy reflects the organization's ability to perform more work in less time and with fewer resources. Organizational synergy is a measure of how well the employees cooperate with one another. Does the organization have a cooperative or noncooperative culture? Cooperative cultures allow for the design of a flexible methodology that will take advantage of continuous improvement opportunities.

Because market conditions and the environment can change, continuous improvement is necessary to maintain the sustained competitive advantage. Change generates risk that, if not properly analyzed and mitigated, can cause a firm to lose its competitive advantage. The key here is for the competitive advantage to become a sustainable competitive advantage. Typical risks associated with maintaining a sustainable advantage are shown in Figure 10.3.

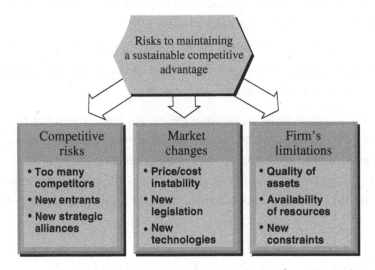

Figure 10.3 Risks associated with maintaining a sustainable competitive advantage.

▶ The Need for Continuous Improvement

Sustained competitive advantages require continuous improvement for a firm to maintain its strength in the marketplace. Although new products/services are one way, strengthening your internal position can also be effective if it results in the introduction of new and/or more sophisticated tools that allow a firm to make faster and better decisions. Tools for the future can be classified as follows:

Resource analysis tools

- Resource-limited planning
- Resource leveling
- Capacity planning
- Multiproject resource analysis

Cost analysis tools

- Earned value forecasting
- Variance analysis
- Trend analysis
- Crashing (or schedule compression) costs

Risk analysis tools

- Risk analysis
- Risk quantification
- Lessons-learned databases

Forecasting analysis tools

- Technology forecasting
- Forward pricing rates
- Escalation factors
- Market analysis

▶ Project Management Competitiveness

Figure 10.4 shows the process of developing project management competitiveness. These steps are somewhat similar to the steps in the project management maturity model (PMMM). In the first step in Figure 10.4, the organization undergoes project management training, which leads to the development of project management skills. But even with a reasonable skill base, the organization can still be reasonably immature. The project management skill base must be regarded as a company-wide project management competency designed to benefit the entire company.

This is more than simply obtaining knowledge. It also includes developing a corporate culture that is based on effective organizational behavior and creating a well-developed project management methodology, accompanied by the proper supporting tools.

Once the organization recognizes that project management is a core competency, the organization can convert this competency into a sustainable competitive advantage, as shown in Figure 10.4. The ultimate purpose is for the sustainable competitive advantage to become the pathway for a strategic competency that becomes a primary effort during strategic planning activities. This requires strong executive support and a firm belief that project management does, in fact, impact the bottom line of the corporation.

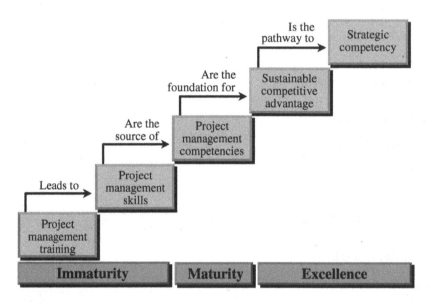

Figure 10.4   Project management competitiveness.

▶ Products versus Solutions

Quite often the need for strategic planning for project management is driven externally by changing customer requirements and expectations rather than internally. As an example, customers are now requesting that contractors provide them with solutions to their business needs rather than products. When you provide a product to a customer, the customer has the responsibility to unpack the project, inspect it, install it, test it, and get it up and running. When you sell a solution to the customer, then you perform all of those activities for the customer and provide the customer with a totally workable product.

To sell solutions to a customer, you must sell not only the product but also the delivery system that will achieve the solution. The delivery system is your project management methodology. You must convince the customer that you have the project management methodology to deliver a solution.

Figure 10.5 illustrates how strategic planning for project management now focuses on solutions. Strategic planning for project management has led to the belief that project management provides the company with a vision that project management can lead to a competitive advantage and should be regarded as a strategic competency. To achieve this vision, the company establishes an intermediate mission that focuses on providing solutions rather than just products.

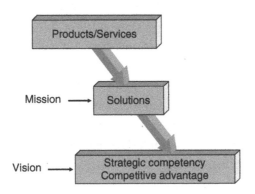

Figure 10.5 Identifying the mission and vision.

Achieving this strategic competency and competitive advantage requires that some foundation elements for project management exist. This is shown in Table 10.1, which identifies what must be done in the near term and long term.

Table 10.1 The foundation elements.

Long term	Short term
Mission	Primary and secondary processes
Results	Methodology
Logistics	Globalization rollout
Structure	Business case development
Accountability	Tools
Direction	Infrastructure
Trust	
Teamwork	
Culture	

▶ Enterprise Project Management

For almost 30 years, project management has meant different things to different people. The focus was always on the end result. Today, the focus is on the delivery system. Historically, every organizational unit within a company was allowed to have its own methodology for producing components, products, and services.

But today, customers want complete solutions, where a *solution* can be defined as the integration of multiple components, products, and services. The result is a requirement for a single methodology, which is called an *enterprise project management approach* and spans the entire company.

Figure 10.6 illustrates the evolutionary process that companies go through to bring in enterprise project management. The vertical arrows reflect functional project management, where each functional area, such as marketing or engineering, could have its own form of project management. This is acceptable as long as customers want products or components. But when a customer wants complete solutions, then all of the functional areas must work together, thus necessitating a single enterprise project management methodology. As companies begin to realize that project management has become a strategic competency, there is a desire for relationships with customers and suppliers to be partnerships rather than just customer-contractor relationships. Therefore, strategic planning on the enterprise project management methodology must occur such that your methodology can easily adapt itself and interface with the methodologies of your customers and suppliers. This is the meaning of the bottom portion of Figure 10.6.

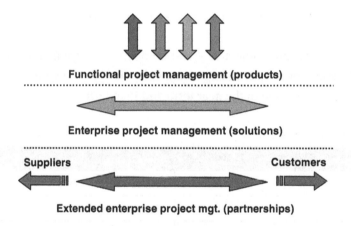

Figure 10.6 Growth of enterprise project management.

► Engagement Project Management

One of the newest terms to enter the project management vocabulary is *engagement project management*. Engagement project management involves the way you approach a prospective client and how you present what you are selling. Given the fact that clients are looking for strategic, long-term partnerships, as well as for solutions rather than just products, you must sell your project management skills to win future business. To provide solutions and a meaningful partnership, engagement project management must emphasize:

- Realistic deliverables and constraints
- An enterprise project management methodology that can deliver continuously and with a high probability of success
- Continuous status reporting
- Customer involvement during decision-making

Engagement project management has created an understanding and demand for project management in both the buyer's and seller's organizations.

CHAPTER ELEVEN

Advanced Project Management Maturity Assessments

▶ Introduction: Changing Times

In today's business environment, a new generation of workers are growing up in a Web 2.0 world of Web-based project management tools that allow people on virtual or distributed teams to work together much more closely than in the past. Advances in computer technology and information flow have shown that the way we traditionally managed projects and defined maturity in the past may have been incorrect and may be ineffective for many of today's projects.

Literature is now appearing describing project management (PM) 2.0 and 3.0, which focus on new project management tools, better project governance, improved collaboration with stakeholders, and more meaningful information reporting using metrics, key performance indicators (KPIs), and dashboard performance reporting systems. The result is that as technology and tools change, so does our need to redefine our definition of project management maturity. Therefore, project management maturity models (PMMMs) need to be updated periodically to account for these changes.

▶ Redefining Maturity from PM 1.0 to PM 2.0/3.0

Project management had its roots in the aerospace, defense, and construction industries more than 50 years ago. Project management practices were effective on large projects with reasonably known and predictable technology, assumptions, and constraints that were unlikely to change over the duration of the project and that were being carried out in a somewhat stable political environment. Unfortunately, for most companies, these types of projects represented only a small portion of all projects they needed to complete to remain in business. Maturity models were required for companies whose projects did not necessarily fit the traditional mold.

Today, we are applying the project management approach to a wider variety of projects encompassing all areas of business; and politics, risk, value, company image and reputation, goodwill, sustainability, and quality are potentially more important to the firm than the traditional time, cost, and scope constraints. As such, the traditional project management practices that we have used for decades, which I will call PM 1.0, are now seen as ineffective for managing some of these new types of projects.

PM 1.0 is based on the following activities:

- Projects are identified, evaluated, and approved without any involvement by project managers.

- Project management is seen as an overhead expense, but a necessity.

- Project planning is done by a centralized planning group, which may or may not include the project manager's input.

- Even though the planners may not fully understand the complexities of the project, the assumption is that the planners can develop the correct baselines and plans, which will remain unchanged for the duration of the project.

- Team members are assigned to the project and expected to perform according to a plan in which they had virtually no input.

- Baselines are established and often approved by senior management without any input from the project team, and again the assumption is that these baselines will not change over the duration of the project.

- Any deviations from the baselines are variances that need to be corrected to maintain the original plan.

- Project success is defined as meeting the planned baselines; resources and tasks may be continuously realigned to maintain the baselines.

- If scope changes are necessary, there is a tendency to approve only those that will not cause the existing baselines to change much.

- Projects are often completed late, over budget, and without satisfying all the customer's requirements.

- If a methodology exists, the accompanying processes, tools, and techniques may not be used in a consistent manner.

Despite all these challenges, companies can still achieve Level 5 maturity accompanied by a continuous stream of successful projects and customer satisfaction, but often with great difficulty in execution. With PM 1.0, executives are fearful that project managers may begin making decisions that should be made only at the executive levels. Senior management wants standardization and control over the way projects are managed, with the expectation that risks will be mitigated and problems will be minimal. Project managers are given very little real authority to make decisions. Almost all business-related and strategic decisions are made by the project sponsors. Enterprise project management (EPM) methodologies were created with the mistaken belief that one size fits all. Every project has to follow the EPM methodology because it supports the executives' comfort zones regardless of the ramifications. The EPM methodologies were constructed around rigid policies and procedures. Project status reporting results in massive reports, and as much as 25 percent of a project's direct labor budget may be consumed by reporting requirements.

Even though a new edition of the *PMBOK® Guide* comes out every four or five years with changes to move us further from PM 1.0, the *PMBOK® Guide* still contains many of the elements of PM 1.0. It may not be possible, or even practical, to create a single *PMBOK® Guide* that can satisfy those firms that still prefer PM 1.0 and those that require the new approaches in PM 2.0 and/or PM 3.0.

▶ Some Critical Issues with PM 1.0

PM 1.0 has worked well for many companies for the types of projects they traditionally managed. But for other companies, PM 1.0 had significant defects that needed to be changed. As an example, conventional project and even business planning, as used with PM 1.0, works on the expectation that managers can predict future outcomes by extrapolating from past results. Planning is often based on history. But for many new business opportunities and forthcoming projects, this way of planning is not possible. Experience may be lacking, or extrapolating from experience may be misleading.

A solution to this problem using PM 2.0 and possibly PM 3.0 is to predict future outcomes based on realistic assumptions. Some of the assumptions made during the planning process will very likely come true, whereas the outcome of others may impact the project to a point that the project should be redirected or even canceled. Project managers may have to test all the assumptions by developing contingency plans based on "what-if" scenarios. However, with PM 1.0, the assumptions that appear in the business case or the project charter are taken as fact and often never challenged. This results in a waste of valuable resources.

Several other PM 1.0 issues needed to be corrected with PM 2.0 and PM 3.0. They included the following:

- Believing that one project management methodology can be applied to every project and will solve almost all problems

- Taking for granted that the constraints and assumptions in the business case/charter are correct and need not be tracked

- Trusting that the planning of others, such as a planning department, is always correct and need not be challenged

- Not having enough metrics to perform trade-off analyses on all problems

- Lacking ownership of plans in which the project manager did not participate, resulting in lack of commitment to the project

- Working with a structured project plan that does not allow for the creativity of team members and possibly trade-offs

---

PMBOK is a registered mark of the Project Management Institute, Inc.

- Not having all necessary information available to the project team

- Working with sponsors and governance committees that do not understand their roles and responsibilities

- Trusting that all decisions made by the sponsors or governance committees are correct

- Believing that implementing project management by executive decree will make it work

- Having no project management culture in the firm

- Believing that a changeover to a project management culture can happen overnight

- Having project management recognized as a part-time addition to a primary job rather than seen as a career path opportunity

- Not understanding the need for project health checks or how to perform a health check

- Having limited tools to support project management activities

- Having too many projects and not enough qualified resources

- Wasting time on projects that need resources we do not have

- Not having any optimization of resources during staffing activities

- Having no benefit-realization plan

- Not understanding how to track benefits or value

- Not working on the projects with the highest business value

- Not recognizing the relationship between the project and strategic business objectives

- Believing that if the project fails, we still have an endless stream of customers

- Not having any meaningful collaboration with stakeholders

- Reporting project information vertically up the organizational hierarchy rather than making information available to the whole team

- Preparing all reports in an optimistic manner with the hope that we can correct any problems before management recognizes the truth

Obviously, there are other issues that could be added to the list. But at least we recognize that a valid need exists for PM 2.0 and PM 3.0, and traditional PMMMs need to be updated to include this material or treat it as an add-on to the PMMM.

▶ The Need for PM 2.0

The idea for PM 2.0 came primarily from project managers involved in software development projects, where adding version numbers to project management seemed necessary because of the different tools being used and different project needs. Over the years, several studies have been conducted to determine the causes of IT project failures. Common failure threads among almost all the studies included lack of user involvement early on, poor governance, and isolated decision-making. These common threads identified the need for distributed collaboration on IT projects, and eventually led to the development of the agile and Scrum approaches. From an IT perspective, we can define PM 2.0 using the following formula:

$$PM\ 2.0 = PM\ 1.0 + distributed\ collaboration$$

Distributed collaboration is driven by open communication. It thrives on collective intelligence that supports better decision-making. Traditional project management favored hierarchical decision-making and formalized reporting, whereas PM 2.0 stresses the need for access to information by the entire project team, including stakeholders and people who sit on the project governance committee.

The need for distributed collaboration is quite clear:

- Stakeholders and members of governance committees are expected to make informed decisions based on evidence and facts rather than making spur-of-the-moment decisions.
- Informed decision-making requires more meaningful metrics.
- The metrics information must be shared rapidly.

Collaboration through formalized reporting can be a very expensive proposition, which is why PM 2.0 focuses heavily on project management metrics, KPIs, and dashboard reporting systems. This increase in collaboration leads some people to believe that PM 2.0 is "socialized project management."

Agile project management is probably today's primary user of PM 2.0. However, there is criticism that the concepts of PM 2.0, accompanied by the heavy usage of distributed collaboration, cannot be used effectively on some large projects. This criticism may have merit. There still exists a valid need for PM 1.0, but at the same time there are attempts to blend together the principles of PM 1.0 and PM 2.0. This creates a challenge when trying to update PMMMs.

All new techniques undergo criticism. PM 2.0 is no exception. Some people argue that PM 2.0 is just a variation of traditional project management. Table 11.1 shows many of the differences between PM 2.0 and PM 1.0. When reading Table 11.1, keep in mind that not all projects, such as those utilizing an agile project management methodology, will necessarily use all the characteristics shown in the PM 2.0 column. This can alter the way assessments in PMMMs are made when considering PM 2.0 characteristics.

Table 11.1 Differences between PM 1.0 and PM 2.0.

Factor	PM 1.0	PM 2.0
Project approval process	Minimal project management involvement	Mandatory project management involvement
Types of projects	Operational	Operational and strategic
Sponsor selection criteria	From funding organization	Business and project management knowledge

(Continued)

Factor	PM 1.0	PM 2.0
Overall project sponsorship	Single-person sponsorship	Committee governance
Planning	Centralized	Decentralized
Project requirements	Well defined	Evolving and flexible
Work breakdown structure (WBS) development	Top down	Bottom up and evolving
Assumptions and constraints	Assumed fixed for duration of project	Revalidated and revised throughout project
Benefit-realization planning	Optional	Mandatory
Number of constraints	Time, cost, and scope	Competing constraints
Definition of success	Time, cost, and scope	Business value created
Importance of project management	Nice-to-have career path	Strategic competency necessary for success
Scope changes	Minimized	Possibly continuous
Activity work flow	In series	In parallel
Project management methodologies	Rigid	Flexible
Overall project flexibility	Minimal	Extensive, as needed
Type of control	Centralized	Decentralized
Type of leadership	Authoritarian	Participative (collaborative)
Overall communications	Localized	Everywhere
Access to information	Localized and restricted	Live, unlimited access, and globalized
Amount of documentation	Extensive	Minimal
Communication media	Reports	Dashboards
Frequency of metrics measurement	Periodically	Continuously
Role of software	As needed	Mandatory
Software tool complexity	Highly complex tools	Easy-to-use tools
Type of contract	Firm fixed price	Cost reimbursable
Responsibility for success	With project manager	With the team
Decision-making	By project manager	By the team
Project health checks	Optional	Mandatory
Type of project team	Co-located	Distributed or virtual
Resource qualifications	Taken for granted	Validated
Team member creativity	Limited	Extensive

Factor	PM 1.0	PM 2.0
Project management culture within firm	Competitive	Cooperative
Access to stakeholders	At selected intervals	Continuous
Stakeholder experience with project management	Optional	Mandatory
Customer involvement	Optional	Mandatory
Organizational project management maturity	Optional	Mandatory
Life-cycle phases	Traditional life-cycle phases	Investment life-cycle phases
Executive's trust in the project manager	Low level of trust	Elevated level of trust
Speed of continuous improvement efforts	Slow	Rapid
Project management education	Nice to have, but not necessary	Necessary and part of life-long learning

Project managers in the future will be given the freedom to select what approach will work best for them on their projects. Rigid methodologies will be replaced by forms, guidelines, templates, and checklists. The project manager will walk through a cafeteria and select from the shelves those elements/activities that best fit a given project. At the end of the cafeteria line, the project manager, accompanied by the project team, will combine all the elements/activities into a project playbook specifically designed for a given client. Client customization will be an essential ingredient of PM 2.0. Rigid methodologies will be converted into flexible methodologies or frameworks.

PM 2.0 is not a separate project management methodology appropriate just for small projects. It is more of a streamlined compilation of many of the updated practices that were embodied in PM 1.0, to allow for a rapid development process. The streamlining was largely due to advances in Web 2.0 software, and success was achieved when everyone on the project team used the same tools.

Although PM 2.0 has been reasonably successful on small projects, the question still exists as to whether PM 1.0 is better for large projects. The jury has not delivered a verdict yet. But some of the publications that discuss how PM 1.0 and PM 2.0 can be combined offer promise. The expected benefits are proactive rather than reactive management, more rapid decision-making, quicker problem resolution, and a better working environment.

▶ The Need for PM 3.0

PM 3.0 focuses heavily on project management as a recognized business process responsible for the delivery of outcomes necessary to meet strategic business objectives.

Heavy emphasis is placed on benefits realization and value-management practices, with the expectation that bad projects either will not make it into the portfolio or will be cancelled early on. Table 11.2 shows some of the differences with PM 3.0.

Table 11.2 Differences between PM 1.0, PM 2.0, and PM 3.0.

Factor	PM 1.0 and PM 2.0	PM 3.0
Project management areas of emphases	Project planning, measuring, and controlling	Benefits realization and value management
Project investment drivers	Cost and profitability	Alignment to strategic business objectives
Metrics selected	To track tangible elements only	To track tangible and intangible elements
Assumptions and constraints	Fixed over the project's life cycle	Can vary over the project's life cycle
Business case development	Unstructured and often with vague assumptions	Structured, including benefits and value identification
Methodologies	Project methodologies; earned value measurement systems (EVMS) and enterprise project management (EPM)	Frameworks and value measurement methodologies (VMM)
Project staffing	Misapplication of critical resources	Capacity planning and resource management

▶ Criticisms of PM 2.0 and PM 3.0

All new techniques bring with them both advantages and disadvantages. This adds significant complexity when deciding to update a PMMM to include PM 2.0 and PM 3.0, because companies may not be using all the activities. The disadvantages of using some of the activities will most certainly lead to criticism. PM 2.0 and PM 3.0 are no exception. Examples of some of the criticisms are:

- PM 2.0 and PM 3.0 are just variations on traditional project management, and the changes would have happened anyway.
- Many companies have track records of success using PM 1.0. Asking them to now use PM 2.0 and PM 3.0 may lead to unnecessary problems.
- PM 2.0 and PM 3.0 work only on IT projects, especially those requiring use of agile or Scrum techniques.
- PM 2.0 advocates open communications, and this may not be possible on large projects. Distribution and control of proprietary information could be an issue as well.
- The data distributed in PM 2.0 may not be auditable, whereas most people believe that PM 1.0 data is auditable.
- Additional tools will have to be created to support PM 2.0 and PM 3.0 implementation. Developing those tools may be expensive.

- Data requirements can easily get out of control, and we may end up with information overload and unwanted executive meddling.
- Although PM 2.0 focuses on collaboration, there is no guarantee that stakeholders or governance committee members will communicate freely with one another.
- PM 2.0 and PM 3.0 will most certainly benefit strategic as well as operational projects, but there is no guarantee that executives will allow project managers to manage strategic projects even if governance is provided.

Another criticism of PM 2.0 and PM 3.0 is having too much information as a result of *metric mania*. There's an old saying: "Be careful what you wish for, because you may get it!" As with any new technique, people often go to extremes rather than following the straight and narrow or simplest path. The real fear with the quest for metrics is that a metric mania mentality will sink in and people will look for the maximum number of metrics that can be collected rather than only what is needed.

While this approach of collecting more metrics than needed may have some merit, the result is usually information overload. The risk is that everyone will want the metrics they found to be permanently part of the metrics database. Not all metrics carry with them an informational value that justifies their use. People may end up collecting too many metrics without fully understanding what the metrics information really means or how it should be used.

When information overload occurs, it may become difficult to identify a core set of metrics for the project. Providing clients and stakeholders with too much or too little information can slow down a project. One of the responsibilities of the project management office's (PMO) policing activities with PM 2.0 and PM 3.0 is ensuring that the correct metrics are placed in the metrics library.

Naysayers will argue against any new technique that may be perceived as pulling them away from their comfort zone. Only time will tell if the criticisms have any merit. But one thing is certain: PM 2.0 and PM 3.0 are being implemented, and they work.

▶ Implementing Continuous Improvement Changes

It is wishful thinking to believe that all the PM 2.0 and PM 3.0 activities listed in Tables 11.1 and 11.2 will evolve naturally. Some of the changes may be initiated by senior management, others by functional management or as a result of client-requested actions, but most of them will be the result of project management continuous improvement initiatives. Someone must assume responsibility for policing the changeover from PM 1.0 to PM 2.0 and then to PM 3.0, and making sure the transition goes smoothly. This is often referred to as transformational project management (TPM). Regardless of the assessment results of the PMMMs, and without some sort of structure and guidance, the continuous improvement initiatives can take much longer than necessary, which will then prolong the time needed to see the benefits of the changes. The policing function should be performed by the PMO.

Traditionally, PMOs were created to help promote the installation and growth of project management. This was particularly true for PM 2.0 implementations.

Implementing or using a PMO included creating a project management methodology and the accompanying forms, guidelines, templates, and checklists. It also included the establishment of a metrics library. As the number of project successes increased, management began assigning additional responsibilities to the PMO. Some of these responsibilities are:

- Forms for standardization in estimating
- Forms for standardization in planning
- Forms for standardization in scheduling
- Forms for standardization in control
- Forms for standardization in reporting
- Clarification of the project manager's roles and responsibilities
- Preparation of job descriptions for project managers
- Preparation of archive data on lessons learned
- Continuous project management benchmarking
- Developing project management templates
- Developing the project management methodology
- Recommending and implementing changes and improvements to the existing project management methodology
- Identifying project management standards
- Identifying best practices in project management
- Performing strategic planning for project management
- Establishing a project management problem-solving hotline
- Coordinating and/or conducting project management training programs
- Transferring knowledge through coaching and mentorship
- Developing a corporate resource capacity/utilization plan
- Assessing risks in projects
- Planning for disaster recovery in projects
- Performing or participating in the portfolio management of projects
- Acting as the guardian for project management intellectual property

Companies began recognizing the return on investment of using a PMO. It is therefore a natural follow-on for the PMO to take the lead with PM 3.0 implementation activities. However, there are significant challenges. Perhaps the greatest challenge is that PM 3.0 focuses on alignment to strategic business objectives as well as the operational objectives most commonly used with PM 1.0. The PMO must now monitor closely how PM 3.0 will interface with all business units rather than just those functional areas that are using project management.

Some companies have established a portfolio PMO that is dedicated to the maximization of benefits and value from the projects within the project portfolio. This requires the use of additional metrics, mainly business-related metrics, needed to answer critical portfolio PMO questions such as these:

- Do we have any weak investments that need to be cancelled or replaced?
- Must any of the programs and/or projects be consolidated?
- Must any of the projects be accelerated or decelerated?
- How well are we aligned to strategic objectives?
- Does the portfolio have to be rebalanced?
- Can we verify that value is being created?
- Do we understand all the risks, and are they being mitigated?
- Are there any indications that we must intervene in some projects?
- Is the information sufficient to predict future corporate performance?
- Must we perform resource reoptimization?

The answer to these questions requires more metrics than most project managers are accustomed to using. These additional metrics are related to:

- Business profitability
- Portfolio health
- Portfolio benefits realization
- Portfolio value achieved
- Portfolio selection and mix
- Resource availability
- Capacity utilization
- Strategic alignment
- Business performance

▶ How to Update the Assessment Instruments

All PMMMs need to be updated. The updates must address issues such as how often to update the material, how much material should be updated, and how to structure the updates. Updates are necessary when the standards for the assessments are updated or when an innovative approach to project management enters the marketplace, such as agile and Scrum. There are three ways to structure the updates to the assessment instruments:

- *Option 1:* Keep the size of the assessment instruments the same, but replace old questions with new questions related to PM 2.0 and PM 3.0.
- *Option 2:* Expand the size of the assessment instrument by including in each section questions related to PM 2.0 and PM 3.0.
- *Option 3:* Update the questions in the traditional assessment where needed, and add new sections that may be optional and contain questions solely related to PM 2.0 and PM 3.0.

Most of the PMMMs in the marketplace are still valid if they are periodically updated to the standards they measure against, such as new editions of the *PMBOK® Guide* and changes in the way the firm defines project management maturity. However, there are risks in the way the updates are made. With option 1, the original assessment instrument may still be valid for companies just starting out in project management. Rewording questions to include advanced levels may eliminate important questions for some firms and scare others away by showing them that they are far from achieving some degree of maturity.

Option 2 has some of the same risks as option 1. Some companies may believe that PM 2.0 and PM 3.0 are attributes solely of agile and Scrum projects, and therefore may not believe it is appropriate to be assessed using these questions. Also, there are companies that are quite good at project management and may use a small percentage of the attributes of PM 2.0 and PM 3.0.

I believe option 3 is best because it allows for customization. With this option, the traditional assessment instruments can be updated and customized according to changes in standards and the definition of maturity. This also has the advantage of allowing existing databases to remain intact for benchmarking purposes. The PM 2.0 and PM 3.0 assessments, because they are separated, can likewise be customized based on how much of PM 2.0 and PM 3.0 the firm uses or plans to use. The risk with this option is that the size of the assessment instruments will increase, and it may take more time to complete the assessments.

► Changing Definitions for PM 2.0 and PM 3.0

The change to PM 2.0 and PM 3.0 is causing some companies to rethink their definition of *project* and *project success*. This is shown in Table 11.3.

Table 11.3 Changing definitions.

Definitions	PM 1.0	PM 2.0 and PM 3.0
Project	According to the *PMBOK® Guide*, 6th edition glossary: "a temporary endeavor undertaken to create a unique product, service or result"	A collection of sustainable business value scheduled for realization
Project success	Completion of the project within the triple constraints of time, cost, and scope	Achieving the desired business value within the competing constraints
Project management maturity	A continuous stream of successful projects	Maximization of portfolio business value that is aligned to strategic objectives

The definitions of a project and project success may change with PM 2.0 and PM 3.0 because of the importance of business benefits realization, business value, the need to align projects to strategic objectives, new project management tools, and other continuous improvement efforts. Some companies can use PM 1.0 activities and still have a continuous stream of successful projects even though they may not be performing in the most efficient or effective manner and may not be utilizing the assigned project resources effectively.

In the future, after PM 2.0 and PM 3.0 practices become everyday occurrences, perhaps we can establish different maturity-assessment instruments for PM 1.0, 2.0, and 3.0. But as of today, perhaps the best we can do is to identify whether a company is using PM 2.0 and PM 3.0 tools and practices to achieve their level of maturity.

▶ Assessing Maturity for PM 2.0 and PM 3.0

The following 10 statements will allow you to determine whether your company is focusing on PM 1.0, PM 2.0, or possibly PM 3.0. Please pick one and only one statement in each of the 10 categories, which represents how your company views metrics. Please read all the statements before selecting.

▶ Statements

1. Metrics:
   A. My company has been using the same traditional metrics for at least the past five years.
   B. My company understands that project performance measurement requires more metrics than just time, cost, and scope.
   C. We are designing processes and templates that include metrics other than time, cost, and scope.
   D. Our methodology currently measures and tracks metrics other than time, cost, and scope.
   E. We currently benchmark against other companies to see what metrics they use other than time, cost, and scope, and how those metrics are being measured.
   F. We are performing continuous improvement efforts on all metrics and have established a metrics library.

2. Project business value:
   A. My company focuses more on completing a deliverable than on the business value that is created.
   B. My company understands that the focus of a project is to create business value rather than just meeting time, cost, and scope constraints.
   C. Our business-case template requires an identification of benefits to be realized and business value being created.
   D. Our methodology tracks metrics for reporting business value created.

    E. We benchmark against other companies as to how they measure and report business value.

    F. We have business-value metrics in our metrics library, and they are updated periodically by our metric owners.

3. Definition of project success:

    A. My company defines project success according to time, cost, and scope.

    B. My company understands that project success requires an understanding of more than just time, cost, and scope.

    C. Our templates and processes identify success criteria using several metrics including business-value metrics.

    D. Our methodology identifies, measures, tracks, and reports business value.

    E. We benchmark against other companies as to how they define project success.

    F. Our best-practices library includes continuous improvement efforts on updating our project-success criteria templates.

4. Project health checks:

    A. My company has no standards in place for performing project health checks.

    B. My company understands the need to periodically perform project health checks.

    C. We are creating templates that should be used for project health checks.

    D. Our methodology identifies core metrics for project health checks.

    E. We benchmark against other companies on how and when they perform project health checks.

    F. Our best-practices library contains data on health check information best practices.

5. Assumptions and constraints tracking:

    A. My company does not track assumptions and constraints as the project progresses.

    B. My company understands the need to track assumptions and constraints the same way we track traditional metrics of time, cost, and scope.

    C. We are developing or have developed templates and processes for tracking assumptions and constraints.

    D. Our methodology allows for the tracking of assumptions and constraints other than time, cost, and scope. We can also challenge the assumptions.

    E. We benchmark against how other companies track and report assumptions and constraints.

    F. Our best-practices library has a section for continuous improvements to assumptions and constraints tracking and how we will challenge the assumptions.

6. Sponsorship and committee governance

    A. Almost all projects are governed by a single person acting in the capacity of a project sponsor.

    B. My company understands that some projects require committee governance rather than single person's sponsorship.

    C. We have templates and processes that define the role of committee governance.

    D. Our methodology allows for customization of performance reporting for the individual needs of members of the governance committee.

    E. We benchmark against other companies to see how they are performing committee governance.

    F. At the end of each project, we capture best practices and lessons learned related to committee governance.

7. Evolving project requirements:

    A. My company will not allow projects to begin until the project's scope is completely defined.

    B. My company understands that, on some projects, a complete upfront definition of the requirements may not be possible, and the project must allow for the requirements to evolve as the project progresses.

    C. We have templates and processes describing how to handle evolving requirements.

    D. Our methodology allows for project execution with evolving requirements.

    E. We benchmark against other companies that allow for evolving requirements.

    F. Our best-practices library has information on lessons learned and best practices when dealing with evolving requirements.

8. Flexible methodologies:

    A. We have a rigid project management methodology that all projects must follow.

    B. My company understands that clients want a flexible methodology that can be customized to their business model.

    C. My company has a methodology that is based on templates and processes that are suitable for customization to a client's needs.

    D. Our methodology or methodologies are treated as frameworks, where the project manager customizes the methodology to fit the client's needs.

    E. We benchmark against how other companies use frameworks rather than rigid methodologies.

    F. Our best-practices library includes lessons learned and best practices on how to customize a methodology to a client's needs.

9. Trust:

   A. The amount of trust that executives have in today's project managers is about the same as it was five years ago.

   B. My company recognizes that more trust must be placed in the hands of the project managers regarding making both project- and business-related decisions.

   C. We are updating/streamlining our templates and processes to reflect the additional trust we have in project managers.

   D. Our methodology allows for the project managers to make significantly more decisions.

   E. We benchmark against how other companies have provided more trust to project managers.

   F. We are capturing best practices and lessons learned on what has happened when the project managers are given more trust.

10. Transformational project management (TPM):

   A. Once our projects are completed, the deliverable are handed over to another group for implementation or execution, and the project manager moves on to the next project.

   B. My company understands that, on some projects, the project manager may eventually become the owner of the deliverables and then act as a functional manager.

   C. We have designed forms and templates that account for the possibility of TPM at the end of the project.

   D. Our methodology can be used for projects that eventually fall into the TPM category.

   E. We benchmark against how other companies handle TPM projects.

   F. We are capturing best practices and lessons learned on TPM projects.

■ Grading System

For each of the 10 statements you selected, award yourself points as follows:

- A: 0 points
- B. 1 point
- C: 2 points
- D: 3 points
- E: 4 points
- F: 5 points

Now add up the total points for all 10 statements. Table 11.4 illustrates how the points might be used as an indicator as to whether you are a user of PM 1.0, 2.0, or 3.0.

Table 11.4 Interpretation of the points.

Points	Interpretation
0–5	Your company appears to be heavily entrenched in PM 1.0. You can still achieve a very high level of maturity according to the traditional PMMM and appear comfortable with the way your projects are managed. You may understand that some changes may be necessary in the future, but you appear somewhat reluctant to understand new practices such as PM 2.0 and PM 3.0.
6–15	Your company still appears entrenched in PM 1.0 practices but understands several of the PM 2.0 and PM 3.0 activities. However, not much is being done to implement any of the PM 2.0 and PM 3.0 activities.
16–25	Even though you may have a reasonably elevated level of maturity according to the PMMM, you understand the need for change. Several of the PM 2.0 and PM 3.0 characteristics are already included in your project management methodology, and continuous improvement to PM 2.0 and PM 3.0 is being done slowly.
26–40	Your company has implemented many of the characteristics of PM 2.0 and PM 3.0.
41–50	You company has implemented most of the PM 2.0 and PM 3.0 characteristics.

Although several other topics are included in PM 2.0 and PM 3.0, only 10 topics were selected for the assessment. The topics in the statements are as follows:

- Advances in metric-measurement techniques have made it possible to use more metrics than just time, cost, and scope to execute a project. It is impossible to determine true performance from just time, cost, and scope.

- We are now creating metrics that can track and measure the creation of business value as the project progresses. Why work on a project if the intent is not to create business value?

- The use of a value metric changes our definition of success. Simply stated, *success is when sustainable business value has been created.*

- Project health checks are now a structured process rather than merely a guess using traditional metrics. New metrics may be needed for health checks.

- The chance that changing assumptions and constraints can cause a project to fail means they must be tracked throughout the project life cycle.

- Sponsors may no longer be able to resolve all problems. Committee governance will become a necessity.

- Not all projects have well-defined requirements at project initiation. Scope must be allowed to evolve, such as with agile project management.

- Project management methodologies will become flexible and will be able to be adapted to the client's business model, thus providing a good opportunity for repeat business.

- Executives are exhibiting more trust in project managers by giving them additional responsibilities for both business and project decisions.
- At the end of some projects, the project manager assumes the role of the business owner for the deliverables that were created. The project manager then acts as a functional manager for the implementation of the deliverables.

▶ Measuring Intangible Benefits and Value

For more than five decades, we have focused heavily on earned value measurement techniques stressing the measurements of just time, cost, and scope because they were well understood and the easiest metrics to measure. They were tangible metrics. For many executives, the information provided using earned value measurement techniques reflects short-term thinking where the benefits and value are attributed to a project that has just ended. Project management was believed to be a series of actions and processes to achieve short-term results. This was in line with executive thinking, since most executives have historically been rewarded for meeting budgets and deadlines. But are there long-term benefits and value from using project management?

The measurement of intangibles is not generally part of PM 2.0 or PM 3.0. Some people in academia, myself included, believe this could be the beginning of PM 4.0. For years, we had questions about intangibles that we had difficulty answering:

- Are corporate project management assets tangible, intangible, or both?
- Can we define intangible project management assets?
- Can intangible project management assets be expressed in financial terms and that can impact the corporate balance sheet?
- Can intangible project management assets be measured?
- Are they value-added, and can we establish intangible project management metrics?
- Will they impact the performance of the organization in the future?

Today, we can measure intangible factors as well as tangible factors that impact project performance. Intangible benefits and value are now seen as intangible assets and can be evaluated by accounting personnel to help indicate the health of a company. An intangible asset is non-monetary and without physical substance. Intangibles may be hard to measure, but they are not immeasurable.

The following are often regarded as intangible project management assets and can be measured:

- *Project management governance:* Did we have proper governance, was it effective, and did the governance personnel understand their roles and responsibilities?

- *Project management leadership:* Did the project manager provide effective leadership?

- *Commitment:* Is top management committed to continuous improvements in project management?

- *Lessons learned and best practices:* Did we capture lessons learned and best practices?

- *Knowledge management:* Are the best practices and lessons learned part of our knowledge management system?

- *Intellectual property rights:* Is project management creating patents and other forms of intellectual property rights?

- *Working conditions:* Are the people on the project teams satisfied with the working conditions?

- *Teamwork and trust:* Are the people on the project teams working together as a team, and do they trust one another's decisions?

As seen in this list, intangible assets are more than just goodwill or intellectual property. They also include maximizing human performance. Understanding and measuring intangible asset value improves performance.

Intangibles are now seen as long-term measurements; most companies no longer focus on just short-term results. Companies now recognize that intangibles impact the long-term bottom line of the firm; they are no longer fearful of what the results will show. And companies no longer argue that they lack the capability to measure intangibles.

► Customizing PM 2.0 and PM 3.0 Assessments

If PM 2.0 and PM 3.0 assessments remain as separate attachments, then customization is relatively easy since not all companies will utilize all the features. However, some companies may wish to customize the traditional assessment instruments by including PM 2.0 and PM 3.0 attributes.

Figures 11.1 through 11.10 match each of the earlier 10 questions to the five levels of maturity in the PMMM. As you might expect, the subject matter for each of the 10 questions is a characteristic of PM 2.0 and PM 3.0, but does not necessarily indicate that a company is mature or immature in project management. Because most of these topics are relatively new to most companies today, they may not be considered as criteria for maturity because companies can become mature without these characteristics (based on the nature of their business.) In future versions of the PMMM, these characteristics may become important and may be fully integrated into each of the levels of the PMMM, as shown in the figures.

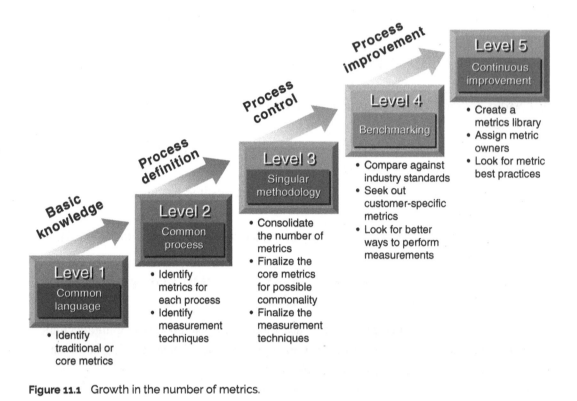

Figure 11.1 Growth in the number of metrics.

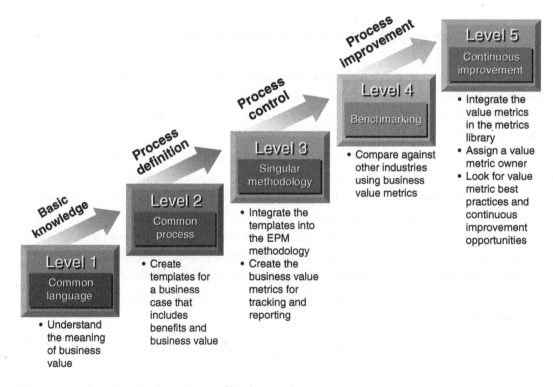

Figure 11.2 Growth in the importance of business value.

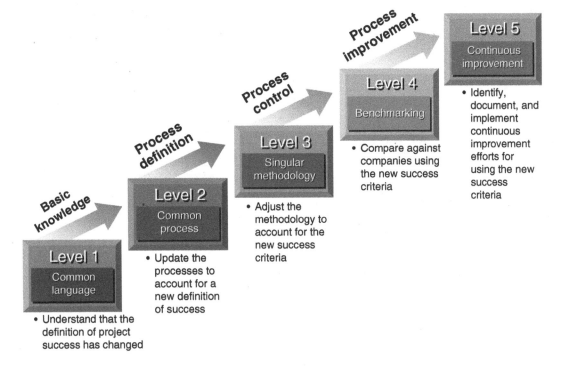

Figure 11.3 Growth in a new definition of project success.

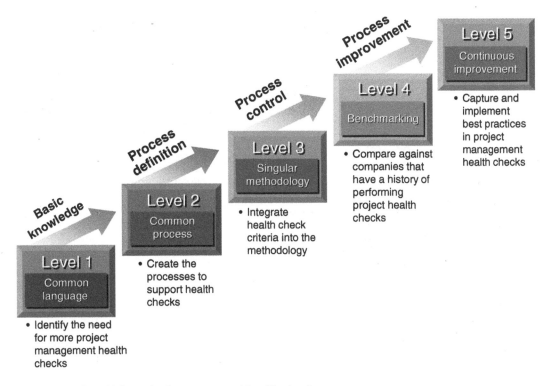

Figure 11.4 Growth in project management health checks.

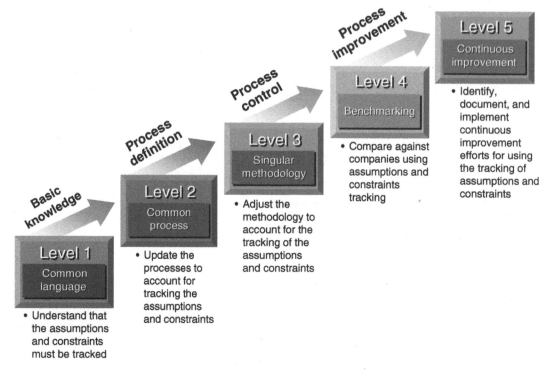

Figure 11.5 Growth in assumptions and constraints tracking.

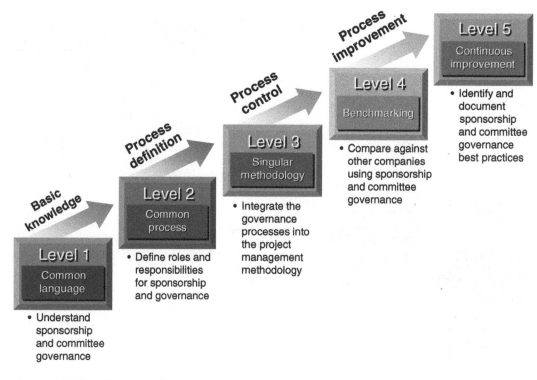

Figure 11.6 Growth in committee governance.

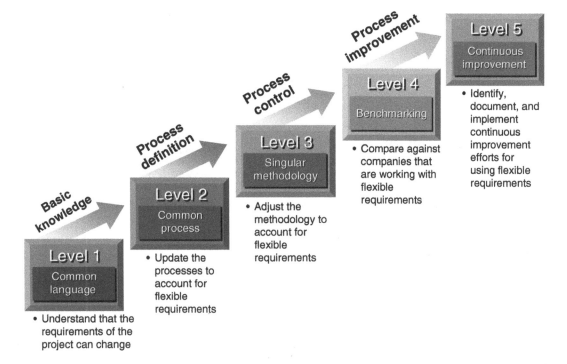

Figure 11.7 Growth in flexible project requirements.

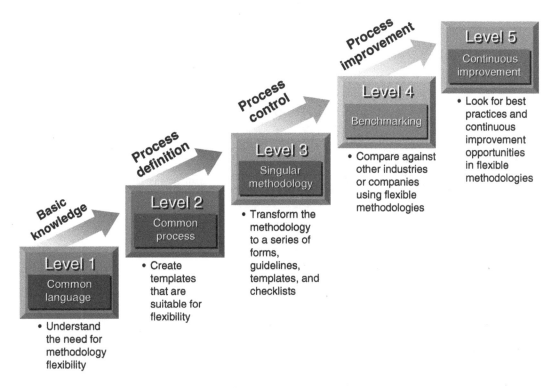

Figure 11.8 Growth in the importance of flexible methodologies (frameworks).

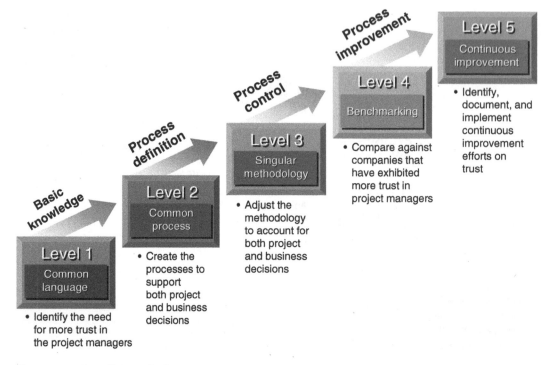

Figure 11.9 Growth in project management trust.

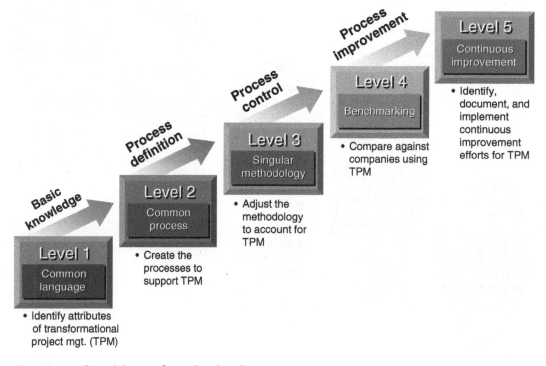

Figure 11.10 Growth in transformational project management.

► PMMM and the Agile Environment[1]

The PMMM discussed in this book has an extensive database comparing the maturity level of hundreds of organizations and thousands of practitioners. It is safe to assume that most of the organizations that used the PMMM in this book had managed projects using a more traditional approach, but it can be customized for agile projects as well.

First, it is important to clarify that a project can be managed and delivered by using a traditional project management approach (i.e., according to the *PMBOK® Guide*) or by using an agile approach (i.e., Scrum). The decision of which approach to use is one that the organization must make based on the nature of the project, the level of uncertainty and complexity, alignment with the overall strategy, the organizational structure, resource availability, and the client's requirements, just to name a few factors.

Related to the assessment, a project management maturity assessment asks questions about how the organization manages its projects; it doesn't matter if the project manager is using a traditional or an agile approach. Specifically, the project management maturity assessment allows people to define how mature an organization is when managing projects, breaking down the results into five distinct levels. In the following, you will see how each level can be applied when using an agile approach.

Level 1 (common language) measures the practitioner's knowledge in project management. When using a traditional way to manage projects, common language is related to the practitioner's knowledge regarding managing projects; understanding of terms like *risk management*, *work breakdown structure*, *earned value management*, *change control system*, *scope statement*, etc.; and, when using agile, understanding of terms like *sprints*, *backlogs*, *burndown charts*, *velocity*, *daily stand-up meetings*, and so on. It can be concluded that a high score on common language means the organization has a great understanding of and knowledge about project management (traditional or agile).

Level 2 (common processes) assesses whether the organization follows common processes across different business units, regions, and departments. Like Level 1, the difference between a traditional approach for delivering projects and the agile approach is that the latter follows agile guidelines (i.e., the Agile Manifesto), with processes followed by the product owner, Scrum Master, and team. With the traditional approach, the project manager and team follow the processes. In both cases, there are roles and responsibilities for the client, sponsors, vendors/suppliers, functional managers, and other stakeholders. If an organization follows common processes when delivering projects, using a traditional or agile approach, it is more mature than another organization that does not. Even in "hybrid" environments—organizations that have a mix of traditional and agile approaches—there will be processes to be followed, and maturity is measured the same way. In other words, processes are processes; it does not matter which approach is being used. More mature organizations will adhere to the defined processes, and less mature organizations will not.

---

[1] Material in this section has been provided by Leon Herszon, Ph.D., CSM, CSPO, PMP, Chief Agility Officer, International Institute for Learning (IIL).

Level 3 (singular methodology) is related to the creation, implementation, and usage of a methodology that is unique to the organization. It is important to note that a project management methodology is not just a collection of processes or steps to deliver a project using the traditional or agile approach. A methodology exists when it incorporates the way an organization does business, considering its values and internal policies and procedures. Even though a methodology can be based on a specific approach (i.e., PRINCE2, *PMBOK Guide®*, Scrum, or SAFe), every methodology is unique. This level measures how mature an organization is related to the existence, understanding, and usage of such a methodology.

Level 4 (benchmarking) is when you can compare your organization's performance delivering projects to others' performance. At this level, it would be useful to have a comparison of agile best practices with organizations that also use agile, so a field in the database may include what approach is being used: traditional, agile, or hybrid. For that reason, it is recommended that benchmarking should allow you to compare your company's performance with others that have the same size, industry, region, or project management approach.

Level 5 (continuous improvement) is focused on the organization's effort to create a lessons-learned database, allowing proper knowledge transfer to future projects. It is also related to the existence of a coaching process that supports practitioners and is aligned with the strategic direction of the organization. The lessons learned from the traditional or agile approaches are used to improve the existing processes, update the methodology, and help improve overall performance. Lessons learned when delivering projects using the agile approach are to be used by professionals working on agile environments. Once again, there is a need to record which approach is being used: agile, traditional, or hybrid.

If the PMMM can select in which environment projects are being managed (traditional, agile, or hybrid), the questions will be properly presented, the answers will be captured, and the results will reflect the maturity level on each of the five levels. The user can also decide to compare the organization's maturity-level results with organizations in the same industry that use a different approach. For instance, a large financial institution, Company A, manages its projects using a traditional approach but would like to compare its maturity results with Company B, which is also a large financial institution but uses agile. The possibilities are endless, but the bottom line is to provide organizations with an action plan that can be put in place to improve their maturity level.

CHAPTER TWELVE

How to Conduct a Project Management Maturity Assessment

▶ Introduction

Once you've decided that a PM maturity assessment is right for your organization, you'll be faced with the somewhat daunting realization that now you've got to actually plan, organize, and implement the assessment. Where do you start? And how do you turn all that assessment data into a meaningful action plan? In this chapter, you will be provided with some helpful guidelines and checklists.

▶ Find Ways to Bypass the Corporate Immune System

Even though what you're doing is for the good of the organization, you may encounter cultural resistance as you prepare to implement a PM maturity assessment. That's because any organization is like a biological organism. It will tend to reject anything that is new and unfamiliar (like the body's immune system rejecting a life-saving transplant). Intellectually the organization will appreciate the value of the assessment, but the company's culture can be a troublemaker. So be sure to take specific steps to prevent "rejection" and to ensure success:

■ Recognize and anticipate that there will be pockets of resistance because people may fear the results will end up removing them from their comfort zone.

■ Acknowledge the fear factor: the apprehension that you may be doing things wrong.

This chapter has been adapted from Harold Kerzner, *Advanced Project Management: Best Practices in Implementation* (New York: Wiley, 2004), 197–208; reproduced by permission of John Wiley & Sons. This chapter was originally prepared by G. Howland Blackiston, formerly executive vice president, International Institute for Learning, Inc. For further information on the Project Management Maturity Assessment Instrument referenced in this chapter, contact Lori Milhaven, Vice President, International Institute for Learning, at 212-515-5121 or via e-mail: lori.milhaven@iil.com.

- Identify the specific cultural issues that might cause resistance. Are there personal issues involved (concern about status, job security, or advancement opportunities)?

- Are there legal restrictions regarding asking employees to take an assessment? Some countries have laws on the books that make assessments difficult.

- Squarely address each and every concern. Defuse resistance by acknowledging problems. Be honest and candid.

- Begin an assessment using volunteers who share your enthusiasm. Let their positive experiences convince the others.

- Launch an assessment effort by beginning with a part of the business that is project driven (e.g., IT, R&D, marketing)

- Start small and scale up. Learn from your early successes and failures before you launch a company-wide assessment effort. Walk before you run.

- Clearly communicate exactly what you are doing and why (see the next section).

▶ Explain Why You Are Doing This

A well-planned and -executed PM maturity assessment is time-consuming. It will involve careful planning and follow-up, and it will draw on considerable resources. So be sure there's clear understanding in the organization as to why you are doing this. Here's where good communication skills come in handy. Prepare a brief and lucid document that can be shared with everyone who will participate in the PM maturity assessment. You've got to sell the importance of doing this assessment. People need to be convinced that this is in both their best interest and the company's. And you've got to disarm resistance. Doing so will help ensure buy-in and head off any problems that can derail success. As you put together this memo, make sure you address the following issues:

- Define what is meant by *project management maturity*.
- Explain why it's important for your company to measure PM maturity.
- Convey how this assessment will make the organization more competitive.
- Underscore how competitive companies create growth and job security.
- Explain who in the company will be invited to participate in the assessment. Why are these people being chosen?
- Describe what is involved and how long it will take.
- Recognize what management will do (and not do) with the assessment information.
- Alleviate any concerns that the information will be used to judge an individual's performance (don't threaten job security).
- Explain how the organization will turn the assessment data into specific improvements.
- Communicate any plans for doing this assessment again (you will want to measure ongoing progress).

Create an Effective Welcome Message

(Here's a sample welcome message from a company using the Kerzner Project Management Maturity Online Assessment Tool. This message appeared on the home page of the web-based assessment tool. Clear statements like this help alleviate cultural concerns about participating in what might be incorrectly regarded as a threatening test of knowledge.)

Welcome to our online project management maturity assessment tool. Project management has been recognized as playing an essential role in our organization. By participating in this assessment you will help create a strategic development plan for identifying the training curriculum that will build and improve our current capabilities. Results of this assessment will be used to set a baseline for all departments and will serve as a tool for identifying future training opportunities. Your support demonstrates your commitment to helping our project management community achieve professional recognition, continuous improvement, and productivity. In turn, this will result in higher consultant/partner/customer satisfaction. We wish to thank you for participating in this groundbreaking event and helping us realize the company's future vision. Thank you for your participation.

—Corporate Project Office

► Pick the Model that Is Best for Your Organization

There's no lack of assessment tools on the market—there are many to choose from—simple or complex, generic or industry specific. There are also numerous formats for assessment reports. An example of an assessment report appears in Appendix A. Basically, they all seek similar objectives: to measure an organization's project management strengths and weaknesses, and to identify improvement opportunities. No one model will be 100 percent perfect for your organization, but some may come close. In all likelihood, you'll wind up with a blend or a customization that best fits your organization. As you evaluate assessment models, consider the following:

- Is it compatible with your project management methodology?
- Does it speak your language (use similar terms and definitions)?
- Has the assessment model been validated (has it been tested and used successfully in other industries)?
- Will this tool work well for your industry? In your organization?
- How easy is the tool to use and administer?
- What delivery mechanism would be best (printed surveys, interviews, online)?
- Is the tool aligned with industry standards (e.g., the *Project Management Book of Knowledge* [*PMBOK*®] *Guide*)?

PMBOK is a registered mark of the Project Management Institute, Inc.

- Global organizations should determine whether the tool is applicable internationally. Customization may be necessary to conform with national laws.

- Can the results of the assessment be easily mapped to your organization's business plan?

- Is the tool flexible? Does it allow for special features and customization?

- Can the tool measure professional skills of project personnel?

- What resources will be required to utilize the tool? How many employees will be involved, and how long will it take?

- What will it cost to undertake the assessment?

▶ Maturity Models: How Do They Compare?

Are you ready to embark on an assessment? In the spirit of comparative analysis, many maturity models are currently available other than the model discussed in this book.

The Origin of Today's Maturity Models

Back in the mid to late 1980s, the software industry explored formal ways to better measure and manage the quality and reliability of the processes used for software development. The industry saw value in applying the concepts of total quality management (TQM) and continuous improvement to their development processes. This prompted the Software Engineering Institute's (SEI) development (in 1990) of the Capability Maturity Model (CMM®). The tool provided the industry with a structured, objective means for measuring a software organization's development processes and comparing these measures against optimum practices. CMM helped software developers identify specific improvements that would allow them to become more competitive in a highly competitive industry. To utilize CMM in other industries, the tools have been blended with project management measures and standards (à la the *PMBOK® Guide*) to serve as the foundation for many of the project management maturity models now on the market.

▶ Create the Right Fit

Some models (like the Kerzner Project Management Maturity Model) have been designed to meet the needs of a broad array of industries and cultures. They are generic. Other models have been developed for specific industries or applications. As you select a model to use in your organization, consider to what degree (if any) the model must be tailored to fit your culture, industry, and business objectives. Some issues to consider are:

- Is the model based on a project management standard that fits with what's used in your organization? Or will you have to tailor the tool to comply with your standards?

- Does the model work in your industry? Are the terms and language used familiar to your business?

- Is there a cultural fit?

- Is the model comprehensive enough to measure leadership, professional development, and management involvement?

- Will the model help you develop a corrective action plan to continuously improve project management processes and practices?

- Does the model allow you to add questions and make modifications without compromising the effectiveness of the assessment?

- Can you sort assessment results to take into account different roles and responsibilities, various departments, geographic locations, or job functions?

It's OK to Make Changes

When tailoring a maturity model to better fit your organization, making changes to the model is perfectly acceptable. For an example, look at our model. Note that Level 3 determines whether the organization is using a singular methodology (rather than using multiple methodologies). Some organizations may intentionally use several methodologies rather than one—for example, one for IT and another for new product development. By all means tailor the model to fit the realities of your organization.

▶ Choose an Appropriate Delivery Method

As you ponder how you will deliver the assessment instrument to your audience, keep in mind that no one way is correct. A number of options are available to you. The method you choose may depend on the audience, size of the company, time available, budget, flexibility, and even technology. Regardless of the option you choose, it's helpful to clarify the time frame for completing the assessment. Let your audience know when you need it back. Don't give them too much time, or they'll put it aside forever. Tell them you need the completed assessment within a couple of weeks. That way, you'll have a better chance of getting what you need in a timely way.

Here are some things to consider:

- Decide if you want an informal or a formal approach. If your organization is small and straightforward, an informal assessment may be all you need.

- If you decide to use conventional questionnaires (paperwork), keep in mind the logistics of distributing, collecting, and tabulating all the data.

- Consider conducting interviews to gather the data you need. This involves some tricky scheduling issues, but if the numbers are manageable, this might be a doable option.

- Don't overlook the possibility of utilizing online technology. This is a very convenient way to reach a large audience in a short time. In addition, the tabulation of results is automatic and instantaneous. And the online format permits easy editing and modification of the tool itself (see sidebar).

- Pick a model and stick with it. Using different instruments may confound your ability to take meaningful corrective action.

- How about using an outside resource (e.g., consultants)? Their objective approach can add value to the assessment results. Often, reports from impartial, outside experts are more readily accepted than the same reports from one of your own.

Use Online Technology to Your Advantage

Creating an interactive web-based assessment has its advantages: easily editable, auto-scoring, efficient distribution, and so on. A few things to consider are covered in the following, along with some screen shots of the Kerzner Project Management Maturity Online Assessment Tool to illustrate how it works. If you decide to spend the time and resources to develop your own web-based assessment, make sure you take into account the following:

- Create an interface that is intuitive and easy to use.
- Make the scoring function automatic.
- Provide for auto-sorting of results by critical filters (key departments, divisions, job functions, hierarchy, etc.).
- Build an Executive Dashboard feature so that the project management office (PMO) and/or top management can continuously monitor the assessment results.

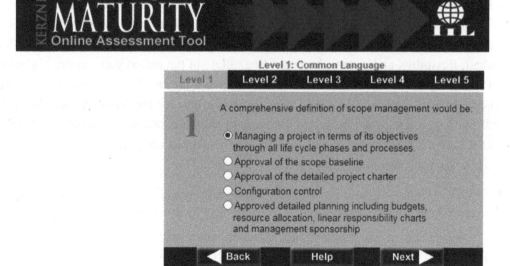

Participants using the Kerzner Assessment Tool answer a series of multiple-choice questions within each of the model's five levels. There are more than 180 questions in the Kerzner Assessment Tool.

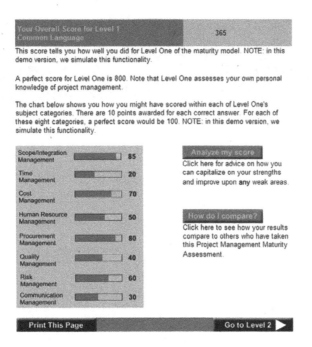

When completing each level of the Kerzner Assessment, participants can see their own score results broken down by the subcategories within that level. This helps them recognize specific strengths and weaknesses.

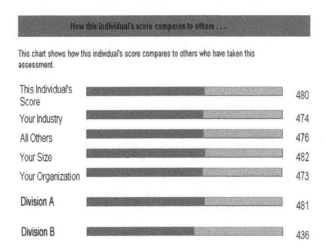

A click of a button compares an individual's score to all others who have taken the Kerzner Assessment—both inside and outside of their organization (a kind of sanity check to see how the results compare to the rest of the world). Users can also see how their company's overall scores compare to other companies in similar industries. Scores are automatically broken down by whatever filtering criteria have been established by the organization. The tool also allows users to display

(continued)

a prescriptive narrative analysis of their results. This auto-generated report suggests what can be done to advance to higher levels of project management maturity (the suggestions are stimulated by the assessment scores).

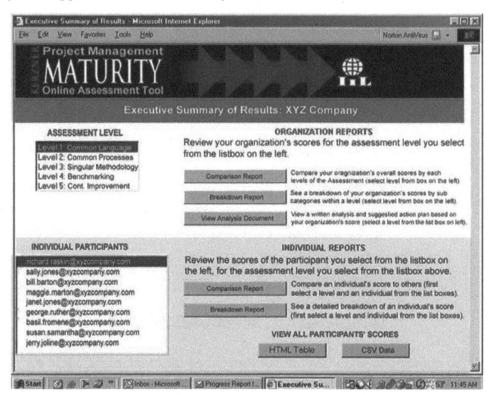

The Kerzner Online Assessment Tool allows authorized managers to see a live summary of their organization's results. Managers can view and compare both individual and company-wide assessment scores. They can even export the raw data into other applications (such as Excel).

▶ Establish Responsibility

In the best of all worlds, the project management office (assuming you have one) should assume overall responsibility for the project management maturity assessment. Chances are, this office reports directly to executive management. Or perhaps its membership consists of executive management. In either case, that link to top management will come in handy, because the assessment will identify opportunities for improvement that will rely on decisions and directives that only they can supply:

- Providing overall leadership for the assessment
- Helping with the selection/design of the assessment tool
- Identifying who should participate in the assessment
- Monitoring assessment results, and aligning the resulting improvement opportunities with the organization's business plan

- Setting priorities (identifying which few actions will have the most meaningful impact)

- Developing an action plan that will allow the organization to achieve these improvements

- Supplying the needed resources (time, budget, personnel)

- Approving the remedial training curriculum

- Encouraging a broadening of the assessment throughout the organization

- Evaluating ongoing progress

▶ Decide Who Should Participate

Here's where you need to make some strategic decisions. Who should participate in the assessment? Ideally, you should include a broad and diverse representation of the entire company—the broader the representation, the more objective and accurate the assessment. Assuming you are attempting to make the company more competitive in project management, you should be assessing the entire organization's project management maturity (not merely assessing the maturity of a single department).

Getting everyone involved may be your ultimate goal, but it may be easier on the company culture to begin in a receptive department before scaling up to include the entire organization. In either case, be certain you have the right people participating in the assessment. Consider the following guidelines as you prepare your invitation list:

- Use the company's business plan as a guide to identify which areas of the organization will critically rely on project management expertise. These departments, locations, or functions are obvious candidates for participating in the assessment.

- Get the right cross-representation involved (the right composition is critical).

- You will underutilize the assessment if you limit participation to experienced project managers (this narrow approach will not accurately reflect the company's actual project management maturity—it's a biased assessment of reality).

- Be sure to include key customers and stakeholders.

- Include broad representation across departments (remember that in world-class companies, all departments, functions, and levels are involved in or knowledgeable about project management efforts).

- Include representation of the entire hierarchy (executive, middle, lower, associates). Experience has shown that each level may have a different belief as to how mature the organization really is.

- Include a large enough sample to be statistically valid. You should invite 30 or more individuals from each operating unit each time you conduct the assessment.

- For large companies, your ultimate objective is to assess at least 10 percent of the population (target larger percentages in small companies).

- Create the ability to sort assessment results by key departments/divisions/job functions/hierarchy (such sorting will enhance your ability to identify project management strengths and weaknesses).

- Decide if participation should be mandatory or optional. For cultural acceptance, I suggest that participation be optional at the start and mandatory as you scale up your assessment efforts and are able to demonstrate value (see "Find Ways to Bypass the Corporate Immune System").

▶ Turn the Results into an Action Plan

With the assessment results in hand, the data should be used to identify meaningful improvement activities. You've got to turn scores into corrective action plans. The prioritization and deployment of these improvements should be spearheaded by the company's PMO. If no such office is in place, consider creating a PMMM assessment action team. Any significant effort will ultimately require the support of top management.

As you organize this effort, keep the following in mind:

- Consider using an objective outside resource to interpret and analyze the assessment results (they've got no axe to grind, and often recommendations from outsiders are more readily accepted by the culture).

- Be as specific as possible when converting assessment data into improvement actions. There's little value in concluding, "Parts of the company need to improve their understanding of risk management." Be much more specific. How is the company going to achieve this goal? A better action would be, "Starting in December, we will schedule two-day intensive workshops in Risk Management for key personnel in our marketing and legal departments."

- Don't forget to identify and deal with any obstacles in the way of making improvements.

- Treat each improvement objective as an individual improvement project (keep in mind that all improvement takes place project by project, and in no other way).

- When prioritizing improvements, start with some home runs (create those bellwether examples that help sell the value of what's being done).

- Focus first on those things that improve the business (prioritize actions based on the organization's business goals and objectives). Use Pareto analysis to identify the vital few improvements that will have the greatest positive impact on the company.

- Use the assessment tool repeatedly to measure progress. Companies that are serious about improving project management should conduct project management maturity assessments at least once every quarter.

► Develop a Remedial Training Curriculum

The assessment will be helpful in identifying a training curriculum that will help you close the gaps. The analysis of scores will clearly identify where training is needed, and in what subjects.

There is great economy here. It means you only need to train those individuals, functions, departments, or locations that have been identified as needing training. It also means you only need training in those specific subjects that have been identified as lacking. This approach subscribes to the concept of "just in time" training, versus "just in case" training.

The PMO (or top management) should establish a task force to plan the organization's remedial training curriculum. Remedial training should be mandated (not voluntary). This group should ensure the following:

- Make certain the assessment results have been analyzed in such a way that it's clear what must be done to make meaningful improvements. If necessary, hire outside experts to help you with the analysis.

- Establish the criteria to be followed in designing or selecting the curriculum.

- Keep the training curriculum keenly responsive to the subject needs and target audiences identified by the assessment ("just in time" versus "just in case").

- Decide whether to buy training from outside firms or develop your own.

- Prepare local case materials as supplements to training. This keeps the tools and methodologies relevant to your business. It also helps participants understand how to apply their new knowledge to their jobs.

- Adapt interactive exercises to fit your culture and job situations.

► Keep Top Management Informed

The overall purpose of doing a project management maturity assessment is to identify opportunities for making significant improvements in the way you manage projects. In turn, such improvements lead to better project outcomes, lower costs, faster results, higher quality, and greater customer satisfaction. But no significant corrective action is possible without the involvement of top management. They are the only ones who can authorize the significant time and resources needed to turn the assessment results into a specific action plan. Keeping top management informed is vital if you are to win their support and leadership.

Part of your assessment plan should include a means for keeping top management informed. One company has 100 employees participate in a project management maturity assessment every month. Every month, top management is given a report summarizing the assessment results. This flow of information helps management prioritize action plans, identify training needs, and measure the company's improvement progress.

Keep the reporting relevant to top management's needs. Share information that is most meaningful to them. This will vary from company to company, but you can be sure that anything expressed in the "language of money" will get their attention and stimulate action. Here are some suggestions:

- Present a detailed breakdown of scores to clearly identify the company's specific strengths and weaknesses.

- Show a comparison of scores between different departments, job functions, geographic locations, or whatever filtering criteria are most meaningful to your business objectives.

- You may wish to provide management with a breakdown of scores by individuals participating in the assessment. This information can be used to identify outstanding talent. And it can even be used to identify individuals who would benefit from remedial training. But be careful! If this information is used to reprimand, criticize, or clean house, you will effectively crush cultural acceptance of the assessment, and all will come to a grinding and hopeless halt.

- Prepare an action plan based on the assessment results. Be specific as to what corrective action is needed and why (see the section "Do It Again").

▶ Virtual Reporting

For those utilizing an online assessment tool, a helpful option is to create an online reporting feature. I refer to this as an Executive Dashboard. It consists of a unique URL address that allows authorized managers to see a detailed summary of their organization's assessment results. Because it's online, the information is real-time, displaying the latest up-to-date scores each time it is accessed. Instant gratification! Managers can view and compare individual and company-wide assessment scores whenever they wish. The feature can also allow them to export the raw data into spreadsheet applications (for other reports, sorting options, etc.). Keep the interface simple and intuitive to use.

▶ Benchmark Your Results to Others

It's helpful to compare your results with those of others who have taken the same assessment (note that the online version of the Kerzner PM Maturity Assessment has this feature). Such comparisons should be both within and outside of your industry. Benchmarking results is helpful for the following reasons:

- Provides a sanity check within your industry (is your maturity level close to your competitor's?).

- Gives you a realistic target, proving that achieving higher levels of maturity is possible (after all, others have already reached higher levels).

- Avoids the deadly sin of resting on your laurels (if you are complacent, assuming you're already best within your industry, it's a sobering jolt to see that other industries are actually much better than you are).

- Sells the need and urgency for improvement (management will be motivated to action if they see that other companies are outperforming your organization).

▶ Do It Again

As helpful as a maturity assessment can be, its usefulness is minimized when it's regarded as a one-time event. That's an underutilization of a powerful tool. Sure, you'll be able to identify strengths, weaknesses, and opportunities for improvement that will help you develop a corrective action plan. But it's only when you use the assessment on a repetitive basis that you can objectively measure the progress of the corrective action plan:

- Are your overall scores improving?

- Is the company achieving higher levels of project management maturity?

- How do you compare to your competition?

- Do you need to modify your corrective action plan, based on the latest assessment results?

- Have new opportunities for improvement emerged since the last assessment?

- Can you improve the assessment tool for more effective use in your organization?

- If one division is outperforming others, are there skills and methods within that exemplary division that can be applied elsewhere in the organization?

Stay nimble and in tune with the marketplace by conducting the assessment on a regular basis. Ongoing use of the tool also allows you to evaluate a larger and larger percentage of your total population. How often should you conduct the assessment? That depends on your organization. Here are some guidelines:

- If your organization is project driven (projects are critical to your business success), you should perform an assessment every month. Vary the audience each time, striving to capture a broad and diverse representation of the organization.

- Other organizations should conduct the assessment a minimum of once a quarter to ensure that improvements are being made. Again, vary the audience each time, striving to capture a broad and diverse representation of the organization.

- Always keep top management informed.

CHAPTER THIRTEEN

Using the PMMM to Extract Best Practices

► Introduction

The ultimate purpose of using a PMMM is continuous improvement efforts. Unfortunately, for many companies the time required for continuous improvement efforts is quite long, and the improvement are done in small increments. By focusing on capturing best practices and lessons learned as part of the assessment, the speed by which continuous improvement efforts are implemented can be accelerated.

As the relative importance of project management permeates each facet of the business, knowledge needs to be captured regarding best practices in project management as well as other business processes. PMMM assessments are not restricted to project management–related questions; the assessments can also identify best practices in program management and portfolio management.

Some companies view this knowledge as intellectual property to be closely guarded in their vaults. Others share this knowledge in hope of discovering other best practices. Companies are now performing strategic planning for project management because of the benefits and its contribution to sustainable business value.

Unfortunately, this is easier said than done. One of the reasons for this difficulty, as will be seen later in the chapter, is that companies today are not in agreement about the definition of a *best practice*, nor do they understand that best practices lead to continuous improvement, which in turn leads to the capturing of more best practices. Many companies also do not recognize the value and benefits that can come from best practices.

Today, project managers are capturing best practices in both project management activities and business activities. The reason is simple: companies are now realizing that best practices are intellectual property that encourages companies to perform at higher levels. Best practices lead to added business value, greater benefit realization, and better benefits management activities. Project management and business thinking are no longer separate activities.

► The Best Practices Process

Why capture best practices? The reasons and objectives for doing so might include:

- Continuous improvements (efficiencies, accuracy of estimates, waste reduction, etc.)
- Enhanced reputation
- Winning new business
- Survival of the firm

Survival of the firm has become the most important reason today for capturing best practices. In the last few years, customers have put pressure on contractors in requests for proposals (RFPs) by requesting the following:

- A listing of the number of PMP® credential holders in the company and how many will be assigned to this project
- A demonstration that the contractor has an enterprise project management methodology, whether rigid or flexible, that is acceptable to the customer; or a requirement that the contractor must use some other methodology approved by the customer
- Supporting documentation identifying the contractor's maturity level in project management, possibly using a PMMM for assessments
- A willingness to share lessons learned and best practices discovered on this project and perhaps previous projects for other customers

Recognizing the need for capturing best practices is a lot easier than doing it. Companies are developing processes for identifying, evaluating, storing, and disseminating information on best practices. There are nine best practices activities, as shown in Figure 13.1, and most companies that recognize the value of capturing best practices accomplish all these steps.

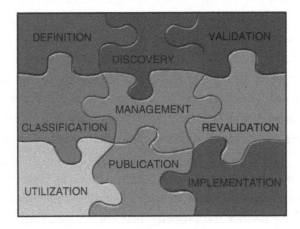

Figure 13.1 Best practices processes.

PMP is a registered mark of the Project Management Institute, Inc.

The processes answer the following nine questions:

- What is the definition of a best practice?
- Who is responsible for identifying the best practice, and where do you look?
- How do you validate that something is a best practice?
- Are there levels or categories of best practices?
- Who is responsible for the administration of the best practice once approved?
- How often do you reevaluate that something is still a best practice?
- How do companies use best practices once they are validated?
- How do large companies make sure everyone knows about the existence of the best practices?
- How do you make sure employees are using the best practices and using them properly?

Each of these questions will be addressed in the next several sections.

▶ Step 1: Definition of a Best Practice

For more than two decades, companies have become fascinated by the expression *best practices*. But now, after two decades or more of use, we are beginning to scrutinize the term, and perhaps better expressions exist.

A best practice begins with an idea that there is a technique, process, method, or activity that can be more effective at delivering an outcome than any other approach and provides the desired outcome with fewer problems and unforeseen complications. As a result, we supposedly end up with the most efficient and effective way of accomplishing a task based on a repeatable process that has been proven over time for a majority of people and/or projects.

But once this idea has been proven to be effective, we normally integrate the best practice into processes so that it becomes a standard way of doing business. Therefore, after acceptance and proven use of the idea, a better expression possibly should be a *proven practice* rather than a *best practice*. This is one argument why *best practice* may be just a buzzword and should be replaced by *proven practice*.

Perhaps in the future the expression *best practices* will be replaced by *proven practices*. For the remainder of this text, I will refer to *best practices*, but please understand that other terms may be more appropriate. This interpretation is necessary here because most of the companies that have contributed to this book still say *best practices*.

Every company can have its own definition of a best practice; there might even be industry standards regarding this definition. Typical definitions of a best practice might be:

- Something that works
- Something that works well
- Something that works well on a repetitive basis

- Something that leads to a competitive advantage
- Something that can be identified in a proposal to generate business
- Something the differentiates you from your competitors
- Something that keeps the company out of trouble and, if trouble occurs, will assist in getting the company out of trouble

There are four primary reasons for capturing best practices:

- Improved efficiency
- Improved effectiveness
- Standardization
- Consistency

▶ Step 2: Seeking Out Best Practices

Best practices can be captured either within your organization or external to your organization. Benchmarking is one way to capture external best practices, possibly by using the project management office (PMO) as the lead for external benchmarking activities. However, there are external sources other than benchmarking for identifying best practices:

- Project Management Institute (PMI®) publications
- Forms, guidelines, templates, and checklists that can affect the execution of the project
- Forms, guidelines, templates, and checklists that can affect your definition of success on a project
- Each of the *PMBOK® Guide* areas of knowledge or domain areas
- Within company-wide or isolated business units
- Seminars and symposiums on general project management concepts
- Seminars and symposiums specializing in project management best practices
- Relationships with other professional societies
- Graduate-level theses

With more universities offering masters- and doctorate-level work in project management, graduate-level theses can provide up-to-date research on best practices.

The problem with external benchmarking is that best practices discovered in one company may not be transferable to another company. In my opinion, most of the best practices are discovered internally and are specifically related to the company's use of its project management methodology, approach, and processes. Good project management methodologies allow for the identification and extraction of best practices. However, opportunities can come from benchmarking as well.

PMBOK is a registered mark of the Project Management Institute, Inc.

Sometimes, the identification of the drivers or metrics that affect each best practice is more apparent than the best practice itself. Metrics and drivers can be treated as early indicators that a best practice may have been found. It is possible to have several drivers for each best practice. It is also possible to establish a universal set of drivers for each best practice, such as:

- Reducing risk by a certain percentage, cost, or time
- Improving estimating accuracy by a certain percentage or dollar value
- Cost savings of a certain percentage or dollar value
- Increasing efficiency by a certain percentage
- Reducing waste, paperwork, or time by a certain percentage

Best practices may not be transferable from company to company, nor will they always be transferable from division to division within the same company. Care must be taken during benchmarking activities to make sure that whatever best practices are discovered are in fact directly applicable to your company.

▶ Step 3: Validating the Best Practice

Seeking out of a best practice is usually done by the project manager, project team, functional manager, and/or possibly a professional facilitator trained in how to debrief a project team and extract best practices. Best practices may also appear as part of PMMM assessments. Any or all of the people involved must believe that what they have discovered is, in fact, a best practice. When project managers are quite active in a project, emphasis is placed on the project manager for the final decision on what constitutes a best practice.

Once the management of the organization affected initially approves the new best practice, it is forwarded to the PMO or process management for validation and then institutionalization. The PMO may have a separate set of checklists to validate the proposed best practice. The PMO must also determine whether the best practice is company proprietary, because that will determine where it is stored and whether it will be shared with customers.

The best practice may be placed in the company's best practices library or, if appropriate, incorporated directly into the company's stage-gate checklist. Based on the complexity of the company's stage-gate checklist process and enterprise project management methodology, the incorporation process may occur immediately or on a quarterly basis.

Some organizations have committees not affiliated with the PMO that have as their primary function the evaluation of potential best practices. Anyone in the company can provide potential best practices data to the committee, and the committee in turn does the analysis. Project managers may be members of the committee. Other organizations use the PMO to perform this work. These committees and the PMO most often report to the senior levels of management.

Evaluating whether something is a best practice is not time-consuming, but it is complex. Simply because someone believes that what they are doing is a best practice does not mean that it is in fact a best practice. Some PMOs are currently developing templates and criteria for determining that an activity may qualify as a best practice. Some items included in the template might be as follows:

- Is transferable to many projects
- Enables efficient and effective performance that can be measured (i.e., can serve as a metric)
- Enables measurement of possible profitability using the best practice
- Allows an activity to be completed in less time and at a lower cost
- Adds value to both the company and the client
- Can differentiate you from everyone else

One company had two unique characteristics in its best practices template:

- Helps to avoid failure
- If a crisis exists, helps us to get out of a critical situation

Executives must realize that these best practices are, in fact, intellectual property that benefits the entire organization. If the best practice can be quantified, then it is usually easier to convince senior management of its value.

▶ Step 4: Levels of Best Practices

Companies that maintain best practices libraries that contain a large number of best practices may create levels of best practices. Figure 13.2 shows various levels of best practices. Each level can have categories within the level. The bottom level is the professional standards level, which would include professional standards as defined by PMI®. The professional standards level contains the greatest number of best practices, but they are more of a general nature than specific and have a low level of complexity.

Figure 13.2 Levels of best practices.

The industry standards level would identify best practices related to performance within the industry. The automotive industry has established standards and best practices specific to the auto industry.

As we progress to the individual best practices in Figure 13.2, the complexity of the best practices goes from general to very specific applications and, as expected, the quantity of best practices is less. An example of a best practice at each level might be (from general to specific):

- *Professional standards:* Preparation and use of a risk management plan, including templates, guidelines, forms, and checklists for risk management.

- *Industry specific:* The risk management plan includes industry best practices such as the best way to transition from engineering to manufacturing.

- *Company specific:* The risk management plan identifies the roles and interactions of engineering, manufacturing, and quality assurance groups during transition.

- *Project specific:* The risk management plan identifies the roles and interactions of affected groups as they relate to a specific product/service for a customer.

- *Individual:* The risk management plan identifies the roles and interactions of affected groups based on their personal tolerance for risk, possibly using a responsibility assignment matrix prepared by the project manager.

Best practices can be extremely useful during strategic planning activities. As shown in Figure 13.3, the bottom two levels may be more useful for project management strategy formulation whereas the top two levels are more appropriate for the execution or implementation of a strategy.

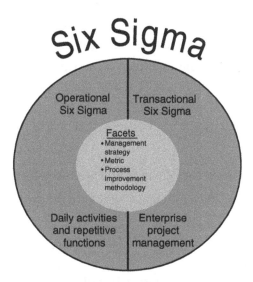

Figure 13.3 Six Sigma categories (nontraditional view).

▶ Step 5: Management of Best Practices

Three players are involved in the management of best practices:

- The best practice's owner
- The PMO
- The best practices library administrator, who may reside in the PMO

The best practice's owner, who usually resides in the functional area, has the responsibility of maintaining the integrity of the best practice. Being a best practice owner is usually an uncompensated, unofficial title but is a symbol of prestige. Therefore, the owner of the best practice tries to enhance it and keep the best practice alive as long as possible.

The PMO usually has the final authority over best practices and makes the final decision about where to place the best practice, who should be allowed to see it, how often it should be reviewed or revalidated, and when it should be removed from service.

The library administrator is merely the caretaker of the best practice and may keep track of how often people review it, assuming it is readily accessible in the best practices library. The library administrator may not have a good understanding of each of the best practices and may not have any voting rights on when to terminate a best practice.

▶ Step 6: Revalidating Best Practices

Best practices do not remain best practices forever. Because they are directly related to the company's definition of project success, the definition of a best practice can change and age as the definition of success changes. Therefore, best practices must be periodically reviewed. The critical question is, "How often should they be reviewed?" The answer is based on how many best practices are in the library. Some companies maintain just a few best practices, whereas large, multinational companies may have thousands of clients and maintain hundreds of best practices in their libraries. If the company sells products as well as services, then there can be both product-related and process-related best practices in the library.

There are usually three types of decisions that can be made during the review process:

- Keep the best practice as is until the next review process.

- Update the best practice and continue using it until the next review process.

- Retire the best practice from service.

▶ Step 7: What to Do with a Best Practice

Given the definition that a best practice is an activity that leads to a sustained competitive advantage, it is no wonder that some companies have been reluctant to make their best practices known to the general public. Therefore, what should a company do with its best practices, if not publicize them? The most common options available include:

- *Share knowledge internally only:* This is accomplished using the company's intranet to share information with employees. There may be a separate group within the company responsible for control of the information, perhaps even the PMO. Not all best practices are available to every employee. Some best practices may be password protected, as discussed next.

- *Hide best practices from all but a select few:* Some companies spend vast amounts of money on the preparation of forms, guidelines, templates, and checklists for project management. These documents are viewed as both company-proprietary information and best practices and are provided to only a select few on a need-to-know basis. An example of a "restricted" best practice might be specialized forms and templates for project approval, where information contained within may be company-sensitive financial data or the company's position on profitability and market share.

- *Advertise to the company's customers:* In this approach, companies may develop a best practices brochure to market their achievements and may also maintain an extensive best practices library that is shared with customers after contract award. In this case, best practices are viewed as competitive weapons.

Even though companies collect best practices, not all best practices are shared outside of the company—even during benchmarking studies, where all parties are expected to share information. Students often ask why textbooks do not include more information on detailed best practices such as forms and templates. One company commented as follows to me:

> We must have spent at least $1 million over the last several years developing an extensive template on how to evaluate the risks associated with transitioning a project from engineering to manufacturing. Our company would not be happy giving this template to everyone who wants to purchase a book for $80. Some best practices templates are common knowledge and we would certainly share this information. But we view the transitioning template as proprietary knowledge not to be shared.

▶ Step 8: Communicating Best Practices Across the Company

Knowledge transfer is one of the greatest challenges facing corporations. The larger the corporation, the greater the challenge of knowledge transfer. The situation is further complicated when corporate locations are dispersed over several continents. Without a structured approach for knowledge transfer, corporations can repeat mistakes as well as miss valuable opportunities. Corporate collaboration methods must be developed.

There is no point in capturing best practices unless the workers know about it. The problem is how to communicate this information to the workers, especially in large, multinational companies. Some of the techniques include:

- Websites
- Best practices libraries
- Community of practice
- Newsletters
- E-mailings
- Internal seminars

- Transferring people
- Case studies
- Other techniques

▶ Step 9: Ensuring Usage of the Best Practices

Why go through the complex process of capturing best practices if people are not going to use them? When companies advertise to their clients that they have best practices, it is understood that tracking of the best practices and how they are used must be done. This is normally part of the responsibility of the PMO. The PMO may have the authority to regularly audit projects to ensure the usage of a best practice but may not have the authority to enforce that usage. The PMO may need to seek out assistance from the head of the PMO, the project sponsor, or various stakeholders for enforcement.

When best practices are used as competitive weapons and advertised to potential customers as part of competitive bidding, the marketing and sales force must understand the best practices and explain this usage to the customers. Unlike 10 years ago, the marketing and sales force today has a good understanding of project management and the accompanying best practices.

▶ Common Beliefs

There are several common beliefs concerning best practices that companies have found to be valid. A partial list follows:

- Because best practices can be interrelated, the identification of one best practice can lead to the discovery of another, especially in the same category or level of best practices. Best practices may be self-perpetuating.
- Because of the dependencies that can exist between best practices, it is often easier to identify categories for best practices rather than individual best practices.
- Best practices may not be transferable. What works well for one company may not work for another company.
- Even though some best practices seem simplistic and based on common sense in most companies, the constant reminder and use of these best practices lead to excellence and customer satisfaction.
- Best practices are not limited exclusively to companies in good financial health. Companies that are cash rich can make a $10 million mistake and write it off. But companies that are cash poor are very careful about how they approve projects, monitor performance, and evaluate whether to cancel projects.

There are reasons why best practices can fail or provide unsatisfactory results, including these:

- Lack of stability, clarity, or understanding of the best practice
- Failure to use best practices correctly

- Identifying a best practice that lacks rigor
- Identifying a best practice based on erroneous judgment
- Failing to provide value

► Best Practices Library

With the premise that project management knowledge and best practices are intellectual properties, how does a company retain this information? The solution is usually the creation of a best practices library. Figure 13.4 shows the three levels of best practices that seem most appropriate for storage in a best practices library.

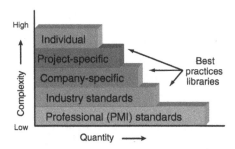

Figure 13.4 Levels of best practices.

Figure 13.5 shows the process of creating a best practices library. The bottom level is the discovery and understanding of what is or is not a potential best practice. The sources for potential best practices can originate anywhere within the organization.

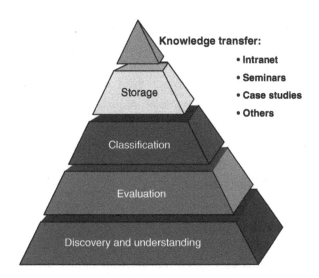

Figure 13.5 Creating a best practices library.

The next level is evaluation, to confirm that it is a best practice. The evaluation process can be done by the PMO or a committee but should have involvement by the senior levels of management. The evaluation process is very difficult because a one-time positive occurrence may not reflect a best practice that will be repetitive. There must exist established criteria for the evaluation of a best practice.

Once a best practice is established, most companies provide a more detailed explanation of it as well as providing a means for answering questions concerning its use. However, each company may have a different approach to disseminating this critical intellectual property. Most companies prefer to make maximum utilization of the company's intranet websites. Other companies consider their current forms and templates to be the ongoing best practices library.

Figure 13.2 showed the levels of best practices, but the classification system for storage purposes can be significantly different. Figure 13.6 shows a typical classification system for a best practices library.

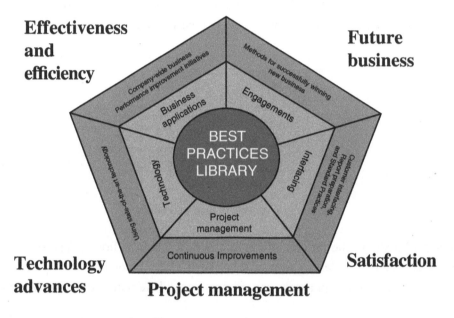

Figure 13.6 Best practices library.

The purpose of creating a best practices library is to transfer knowledge to employees. Knowledge can be transferred through the company intranet, seminars on best practices, and case studies. Some companies require that each project team prepare case studies on lessons learned and best practices before the team is disbanded. These companies then use the case studies in company-sponsored seminars. Best practices and lessons learned must be communicated to the entire organization. The problem is determining how to do so effectively.

Another critical problem is best practices overload. One company started a best practices library and, after a few years, had amassed what it considered to be hundreds of best practices. For a long time, however, nobody bothered to reevaluate whether all of these were still best practices. Once reevaluation finally took place, it was determined that fewer than one-third of the practices listed were still regarded as best practices. Some were no longer best practices, others needed to be updated, and others had to be replaced with newer best practices.

Every company has its own approach regarding when to reevaluate a best practice. Some companies do so quarterly or semiannually, whereas others do so on the anniversary date of the best practice.

The fastest way to achieve some degree of maturity in project management is through the capturing and implementation of best practices. The tricky part is how to capture them. Some companies hold debriefing sessions at the end of each project to discuss what was learned. In these meetings, people seem willing to discuss lessons learned and best practices only from their successes.

More best practices could be learned from failures than successes, but people resist discussing failures if their name can be associated with a possible failure and the result could be held against them during a performance review. Some companies use professional facilitators rather than project managers to hold the debriefing sessions because the facilitators know how to get people to discuss failures as well as successes.

The PMMM assessments are an excellent way to get employees to discuss lessons learned and best practices, provided that the names of individuals are not attached to the assessment instruments. This will require some degree of customization with the intent of seeking out additional information that is not to be evaluated using a point system, because not everyone will be able to discover best practices.

The risk in doing this is that not all the people on each project team will be taking the assessments. Timing is also an issue because most best practices are identified at the end of a project or after a certain amount of time working on a project. You cannot always ask people to participate in the PMMM assessments only when they are near the end of the projects they are working on.

CHAPTER FOURTEEN

Case Studies

▶ Case 1: Simone Engineering Company

Simone Engineering was an engineering component manufacturing firm as well as an engineering consulting company. Simone had an excellent reputation for stakeholder relations management practices, product quality, and customer service.

On all projects above a certain dollar value, sponsorship was performed by the executive levels of management. The executive sponsors were responsible for stakeholder relations management; and, on some projects, only the sponsors handled direct communications with stakeholders and clients. On these projects, project managers met with the stakeholders and clients only during the stakeholder/client interface meetings.

Even though project management existed in the firm and worked reasonably well, most of the business decisions for projects were made by the project sponsors. Project managers often believed that they were simply performing as puppets, even though the company provided project management training courses that followed effective project management practices as stated in the *PMBOK® Guide*.

Although Simone often received single-source and sole-source procurement contracts, much of its business was obtained through competitive bidding. Simone never had any issues responding to a request for proposal (RFP) until now. This client stated in the RFP that all bidders must show how mature they were in project management using a PMMM, and that the report from the PMMM must be no more than six months old.

Simone had never assessed its organization for project management maturity because the business appeared to be doing well. Now a decision had to be made, and there were both pros and cons concerning a PMMM assessment. One of the biggest concerns was fear that the results of the assessment would show that Simone was not performing sponsorship correctly. Additional questions were raised by senior management in deciding whether to perform the PMMM assessments:

- If we do not perform an assessment, will this eliminate us from further competitive bidding with this client?

PMBOK is a registered mark of the Project Management Institute, Inc.

- What happens if other clients have the same requirements in their RFPs? How long can we hold off? How will it affect our business if we are non responsive?

- How will the client react if the assessment results show that we are not very mature?

- Is there a chance the client might cancel the contract downstream if they do not see any improvements toward maturity?

- How quickly can we implement the assessments in a PMMM?

- How many resources should participate in the assessments?

- How do we decide which PMMM to use? Should it be an internal decision, or should we use consultants for assistance in PMMM selection?

- What might happen if the results show that we are doing things wrong? How will our workers react? Will the workers want to see changes?

- If changes are needed, how quickly can the changes be made?

- What might happen if the changes are never implemented?

- Might the workers resist any changes that could remove them from their comfort zones?

- Should we be fearful that the changes needed might reduce the power and authority that senior management now possesses?

- How often must we perform the assessments to validate that changes have been made?

Senior management now had to decide if, and when, they should perform an assessment.

■ Questions

1. Do these seem like realistic questions for senior management to consider?

2. What might happen if these questions are not considered in advance?

► Case 2: NorthStar Software Company

NorthStar is a software consulting company with an excellent reputation for new product development. NorthStar started up more than two decades ago. The company created an IT project management methodology, also referred to as a life-cycle systems development methodology, centered around the *PMBOK® Guide*. During competitive bidding activities, NorthStar promoted in its proposal that "how it delivered the results" was just as important as "what it delivered."

NorthStar established a PMO to continuously look for ways to improve its delivery system. Initially, the PMO asked all project teams at the end of each project to provide to the PMO all the best practices and lessons learned discovered. Unfortunately, some of the projects that were more than a year or two in duration had resources reassigned during the life cycle of the project, and many of the lessons learned and best practices

---

PMBOK is a registered mark of the Project Management Institute, Inc.

were lost. The PMO then changed the policy and asked project teams to provide the PMO with the information at the end of each life-cycle phase, thus getting continuous improvement efforts implemented more quickly.

Attached to the PMO was an informal committee made up of functional employees who were knowledgeable in project management. The informal committee members reported solid to their functional managers but dotted to the PMO. The committee evaluated all lessons learned, both good and bad, to validate their importance as best practices and their applicability to a multitude of projects. Many of the best practices that were finally accepted were placed into the PMO's best practices library to be shared with the rest of the company. Almost all of the best practices were continuous improvements to the company's project management methodology.

Utilizing an IT methodology that followed the *PMBOK® Guide* was important to NorthStar because most of the company's clients used the *PMBOK® Guide* and considered it the project management standard. With advances in agile and Scrum applications to IT projects, NorthStar began updating its project management approach. The company changed from a rather inflexible project management methodology to a flexible methodology or framework that could be customized to each client's needs. NorthStar also eliminated many of the traditional *PMBOK® Guide* processes and activities that were not part of the agile or Scrum approach.

Although converting from a traditional or "waterfall" project management approach to agile and Scrum seemed like a promising idea, it brought with it some headaches when dealing with clients that were unfamiliar with agile and Scrum. NorthStar had to prepare for issues that required demonstrating its project management maturity.

■ Questions

1. If a client were to ask NorthStar how mature the company was in project management, how should NorthStar respond if we assume there are no PMMMs in the marketplace that focus solely on agile and Scrum?

2. If a client were to ask NorthStar to show its level of project management maturity using a traditional PMMM, how should NorthStar respond?

3. Let's assume several PMMMs are available specifically for agile and Scrum approaches. Should NorthStar use one of those models, knowing the clients will want to see the reports?

▶ Case 3: Colmar Automotive

Colmar Automotive is a Detroit-based tier-one supplier to the auto industry. In 2010, Colmar installed a project management methodology based on nine life-cycle phases. For the next four years, all 60,000 employees worldwide accepted the methodology and used it. Management was pleased with the results. Colmar periodically performed an assessment of the maturity of its methodology, and both Colmar and its clients were pleased with the results. Colmar's customer base provided the company with quality recognition awards that everyone believed were attributable to how well the project management methodology was executed.

In February 2016, Colmar Automotive decided to offer additional products to its customers. This strategic move required the purchasing of two European companies. Because the employees from all three companies would be working together closely, a singular project management methodology would be required that would be acceptable to all three companies. One of the European companies had its own methodology based on five life-cycle phases. The second European company had a methodology that was based on rigid policies and procedures and did not look similar to the other two methodologies. The two European methodologies were aligned to European standards rather than the *PMBOK® Guide*. All three methodologies had advantages and disadvantages and appeared to be well liked by their customers.

It took almost two years for all three companies to come to an agreement on a standardized delivery system for projects. Each company made concessions. Now, Colmar was concerned about how to show the maturity of the new agreed-on approach.

■ Questions

1. Can a PMMM be structured so as to consider more than one project management standard, such as the European standard and the *PMBOK® Guide*?

2. How should Colmar respond if the assessments are based on just one standard and the results show that Colmar is not quite mature in project management?

3. How should Colmar respond if the European companies prefer a PMMM based on European standards and American companies prefer an American standard?

4. What if Colmar were to perform the assessments with two PMMMs and they showed different results, such as different levels of maturity and different issues?

► Case 4: Ferris HealthCare, Inc.

In July 2014, senior management at Ferris HealthCare recognized that its future growth could very well be determined by how quickly and how well it implemented project management. For the past several years, line managers had been functioning as project managers while still managing their line groups. Projects came out with the short end of the stick, most often late and over budget, because managers focused on line activities rather than project work. Everyone recognized that project management needed to be an established career-path position and that some structured process had to be implemented for project management.

A consultant was brought into Ferris to provide initial project management training for 50 of the 300 employees targeted for eventual project team assignments. Following the training, several of the employees who attended the training sessions were placed

PMBOK is a registered mark of the Project Management Institute, Inc.

on a committee with senior management to design a project management stage-gate model for Ferris.

After two months of meetings, the committee identified the need for three different stage-gate models: one for information systems, one for new products/services provided, and one for bringing on board new corporate clients. There were several similarities among the three models. However, personal interests dictated the need for three methodologies, all based on rigid policies and procedures.

After a year of using three models, the company recognized it had a problem deciding how to assign the right project manager to the right project. Project managers had to be familiar with all three methodologies. The alternative, considered impractical, was to assign only those project managers familiar with that specific methodology.

After six months of meetings, the company consolidated the three methodologies into a single methodology, focusing more on guidelines than on policies and procedures. The entire organization appeared to support the new singular methodology. A consultant was brought in to conduct the first three days of a four-day training program for employees not yet trained in project management. The fourth day was taught by internal personnel with a focus on how to use the new methodology. The success to failure ratio on projects increased dramatically.

■ Questions

1. If a PMMM assessment was made when Ferris was beginning to implement project management, would the results have shown that Ferris was reasonably mature or immature in its approach toward maturity?

2. Should a PMMM assessment be performed early on, when a company is first beginning its journey toward project management?

► Case 5: Clark Faucet Company

■ Background

By 2010, Clark Faucet Company had grown into the third-largest supplier of faucets for both commercial and home use. Competition was fierce. Consumers evaluated faucets on artistic design and quality. Each faucet had to be available in at least 25 different colors. Commercial buyers seemed more interested in the cost than the average consumer, who viewed the faucet as an object of art, regardless of price.

Clark Faucet Company did not spend a great deal of money advertising on the radio or on television. Some money was allocated for ads in professional journals. Most of Clark's advertising and marketing funds were allocated to two semiannual home-and-garden trade shows and an annual builders trade show. One large builder could purchase more than 5,000 components for furnishing one newly constructed hotel or one apartment complex. Missing an opportunity to display new products at these trade shows could easily result in a 6- to 12-month window of lost revenue.

■ Culture

Clark Faucet had a noncooperative culture. Marketing and engineering never talked to one another. Engineering wanted the freedom to design new products, whereas marketing wanted final approval to make sure that what was designed could be sold.

The conflict between marketing and engineering became so fierce that early attempts to implement project management failed. Nobody wanted to be the project manager. Functional team members refused to attend team meetings and spent most of their time working on their own pet projects rather than the required work. Their line managers also showed little interest in supporting project management.

Project management became so disliked that the procurement manager refused to assign any of his employees to project teams. Instead, he mandated that all project work come through him. He eventually built up a large brick wall around his employees. He claimed that this would protect them from the continuous conflicts between engineering and marketing.

■ The Executive Decision

The executive council mandated that another attempt to implement good project management practices must occur quickly and that PMMM assessments should be made periodically. Project management would be needed not only for new product development but also for specialty products and enhancements. The vice presidents for marketing and engineering reluctantly agreed to try to patch up their differences but did not appear confident that any changes would take place.

Strange as it may seem, nobody could identify the initial cause of the conflicts or how the trouble began. Senior management hired an external consultant to identify the problems, provide recommendations and alternatives, and act as a mediator. The consultant's process began with interviews.

■ Engineering Interviews

The following comments were made during engineering interviews:

- "We are loaded down with work. If marketing would stay out of engineering, we could get our job done."
- "Marketing doesn't understand that there's more work for us to do other than just new product development."
- "Marketing personnel should spend their time at the country club and in bar rooms. This will allow us in engineering to finish our work uninterrupted!"
- "Marketing expects everyone in engineering to stop what they are doing in order to put out marketing fires. I believe that most of the time the problem is that marketing doesn't know what they want up front. This leads to change after change. Why can't we get a good definition at the beginning of each project?"

■ Marketing Interviews

The following comments were made during marketing interviews:

- "Our livelihood rests on income generated from trade shows. Since new product development is four to six months in duration, we have to beat up on engineering to make sure that our marketing schedules are met. Why can't engineering understand the importance of these trade shows?"

- "Because of the time required to develop new products [four to six months], we sometimes have to rush into projects without having a good definition of what is required. When a customer at a trade show gives us an idea for a new product, we rush to get the project underway for introduction at the next trade show. We then go back to the customer and ask for more clarification and/or specifications. Sometimes we must work with the customer for months to get the information we need. I know that this is a problem for engineering, but it cannot be helped."

The consultant wrestled with the comments but was still somewhat perplexed. "Why doesn't engineering understand marketing's problems?" pondered the consultant. In a follow-up interview with an engineering manager, the following comment was made:

"We are currently working on 375 different projects in engineering, and that includes those which marketing requested. Why can't marketing understand our problems?"

■ Questions

1. If a PMMM assessments were conducted, what issues would most likely be discovered?
2. What should be the starting point for continuous improvements?
3. How long should it take to implement the first set of changes?

▶ Case 6: Macon, Inc.

Macon was a 50-year-old company in the business of developing test equipment for the tire industry. The company had a history of segregated departments with very focused functional line managers. The company had two major technical departments: mechanical engineering and electrical engineering. Both departments reported to a vice president for engineering, whose background was always mechanical engineering. For this reason, the company focused all projects from a mechanical engineering perspective. The significance of the test equipment's electrical control system was often minimized, when the electrical control systems were what made Macon's equipment outperform that of the competition.

Because of the strong autonomy of the departments, internal competition existed. Line managers frequently competed with one another rather than focusing on the best interest of Macon. Each hoped the other would be the cause of project delays instead of working together to avoid project delays altogether. Once dates slipped, fingers were pointed, and problems worsened over time.

A typical issue that Macon faced more than once was a customer with a service department that always blamed engineering for its problems. If a machine was not assembled correctly, it was engineering's fault for not documenting it clearly enough. If a component failed, it was engineering's fault for not designing it correctly. No matter what problem occurred in the field, customer service always put the blame on engineering.

As might be expected, engineering blamed most problems on production, claiming that production did not assemble the equipment correctly and did not maintain the proper level of quality. Engineering designed products and then threw them over the fence to production without ever going down to the manufacturing floor to help with assembly. Errors or suggestions reported from production to engineering were ignored. Engineers often perceived the assemblers as incapable of improving designs.

Production ultimately assembled products and shipped them out to customers. Often, during assembly, the production people changed designs as they saw fit without involving engineering. This caused severe problems with documentation. Customer service would later inform engineering that the documentation was incorrect, once again causing conflict among all departments.

The president of Macon was a strong believer in project management. Unfortunately, his preaching fell on deaf ears. The culture was just too strong. Projects failed miserably. Some failures were attributed to the lack of sponsorship or commitment from line managers. One project failed as the result of a project leader who failed to control scope. Each day the project fell further behind because work was added with very little regard for the project's completion date. Project estimates were based on a "gut feel" rather than on sound quantitative data.

The delay in shipping dates led to increasing frustration for customers. Customers began assigning their own project managers as watchdogs to look out for their companies' best interests. The primary function of these watchdog project managers was to ensure that the equipment purchased was delivered on time and complete. This involvement by customers was becoming more noticeable.

The president decided that action was needed to achieve some degree of excellence in project management. The question was what action to take, and when.

■ Questions

1. Does it make sense to perform a PMMM assessment, and, if so, what are the benefits?

2. What are the disadvantages of performing an assessment?

"This is impossible! Just totally impossible! Ten months ago I was sitting on top of the world. Upper-level management considered me one of the best, if not the best, engineer in the plant. Now look at me! I have bags under my eyes, I haven't slept soundly in the last six months, and here I am, cleaning out my desk. I'm sure glad they gave me back my old job in engineering. I guess I could have saved myself a lot of grief and aggravation had I not accepted the promotion to project manager."

■ History

Gary Anderson had accepted a position with Parks Corporation right out of college. With a Ph.D. in mechanical engineering, Gary was ready to solve the world's most traumatic problems. At first, Parks Corporation offered Gary little opportunity to do the pure research that he eagerly wanted to undertake. However, things soon changed. Parks grew into a major electronics and structural design corporation during the big boom of the late 1950s and early 1960s when Department of Defense (DoD) contracts were plentiful.

Parks Corporation grew from a handful of engineers to a major DoD contractor, employing some 6,500 people. During the recession of the late 1960s, money became scarce and major layoffs resulted in lowering the employment level to 2,200 employees. At that time, Parks decided to get out of the research and development (R&D) business and compete as a low-cost production facility while maintaining an engineering organization solely to support production requirements.

After attempts at virtually every project management organizational structure, Parks Corporation selected the matrix form. Each project had a program manager who reported to the director of program management. Each project also maintained an assistant project manager—normally a project engineer—who reported directly to the project manager and indirectly to the director of engineering. The program managers spent most of their time worrying about cost and time, whereas the assistant program managers worried more about technical performance.

With the poor job market for engineers, Gary and his colleagues began taking coursework toward MBA degrees in case the job market deteriorated further. In 1995, with the upturn in DoD spending, Parks had to change its corporate strategy. Parks had spent the last seven years bidding on the production phase of large programs. Now, however, with the new evaluation criteria set forth for contract awards, those companies winning the R&D and qualification phases had a definite edge on being awarded the production contract. The production contract was where the big profits could be found. In keeping with this new strategy, Parks began to beef up its R&D engineering staff. By 1998, Parks had increased in size to 2,700 employees. The increase was mostly in engineering. Experienced R&D personnel were difficult to find for the salaries that Parks was offering. Parks was, however, able to lure some employees away from competitors, but it relied mostly on the younger, inexperienced engineers fresh out of college.

With the adoption of this corporate strategy, Parks Corporation administered a new wage and salary program that included job upgrading. Gary was promoted to senior scientist, responsible for all R&D activities performed in the mechanical engineering department. Gary had distinguished himself as an outstanding production engineer during the past several years, and management felt that his contribution could be extended to R&D as well.

In January 1998, Parks Corporation decided to compete for Phase I of the Blue Spider Project, an R&D effort that, if successful, could lead into a $500 million program spread out over 20 years. The Blue Spider Project was an attempt to improve the structural capabilities of the Spartan missile, a short-range tactical missile used by the Army. The Spartan missile was exhibiting fatigue failure after six years in the field. This was three years less than what the original design specifications called for. The Army wanted new materials that could result in a longer life for the Spartan missile.

Lord Industries was the prime contractor for the Army's Spartan Program. Parks Corporation would be a subcontractor to Lord if it could successfully bid and win the project. The criteria for subcontractor selection were based not only on low bid but also on technical expertise as well as management performance on other projects. Parks's management felt that it had a distinct advantage over most of the other competitors because the company had successfully worked on other projects for Lord Industries.

■ The Blue Spider Project Kickoff

On November 3, 1997, Henry Gable, Parks's director of engineering, called Gary Anderson into his office and said:

> Gary, I've just been notified through the grapevine that Lord will be issuing the request for proposal [RFP] for the Blue Spider Project by the end of this month, with a 30-day response period. I've been waiting a long time for a project like this to come along so that I can experiment with some new ideas that I have. This project is going to be my baby all the way! I want you to head up the proposal team. I think it must be an engineer. I'll make sure that you get a good proposal manager to help you. If we start working now, we can get close to two months of research in before proposal submittal. That will give us a one-month's edge on our competitors.

Gary was pleased to be involved in such an effort. He had absolutely no trouble in getting functional support for the R&D effort necessary to put together a technical proposal. All of the functional managers continually remarked to Gary, "This must be a biggie. The director of engineering has thrown all of his support behind you."

On December 2, the RFP was received. The only trouble area that Gary could see was that the technical specifications stated that all components must be able to operate normally and successfully through a temperature range of –65°F to 145°F. Current testing indicated the Parks Corporation's design would not function above 130°F. An intensive R&D effort was conducted over the next three weeks. Everywhere Gary looked, it appeared that the entire organization was working on his technical proposal.

A week before the final proposal was to be submitted, Gary and Henry Gable met to develop a company position concerning the inability of the preliminary design material to be operated above 130°F.

Gary Anderson: "Henry, I don't think it is going to be possible to meet specification requirements unless we change our design material or incorporate new materials. Everything I've tried indicates we're in trouble."

Henry Gable: "We're in trouble only if the customer knows about it. Let the proposal state that we expect our design to be operative up to 155°F. That'll please the customer."

Anderson: "That seems unethical to me. Why don't we just tell them the truth?"

Gable: "The truth doesn't always win proposals. I picked you to head up this effort because I thought that you'd understand. I could have just as easily selected one of our many moral project managers. I'm considering you for program manager after we win the program. If you're going to pull this conscientious crap on me like the other project managers do, I'll find someone else. Look at it this way; later we can convince the customer to change the specifications. After all, we'll be so far downstream that they'll have no choice."

After two solid months of 16-hour days for Gary, the proposal was submitted. On February 10, 1998, Lord Industries announced that Parks Corporation would be awarded the Blue Spider Project. The contract called for a 10-month effort, negotiated at $2.2 million at a firm-fixed price.

■ Selecting the Project Manager

Following contract award, Henry Gable called Gary in for a conference.

Henry Gable: "Congratulations, Gary! You did a fine job. The Blue Spider Project has great potential for ongoing business over the next 10 years, provided that we perform well during the R&D phase. Obviously you're the most qualified person in the plant to head up the project. How would you feel about a transfer to program management?"

Gary Anderson: "I think it would be a real challenge. I could make maximum use of the MBA degree I earned last year. I've always wanted to be in program management."

Gable: "Having several masters' degrees, or even doctorates for that matter, does not guarantee that you'll be a successful project manager. There are three requirements for effective program management: You must be able to communicate both in writing and orally; you must know how to motivate people; and you must be willing to give up your car pool. The last one is extremely important in that program managers must be totally committed and dedicated to the program, regardless of how much time is involved.

"But this is not the reason why I asked you to come here. Going from project engineer to program management is a big step. There are only two places you can go from program management—up the organization or out the door. I know of very, very few engineers who failed in program management and were permitted to return."

Anderson: "Why is that? If I'm considered to be the best engineer in the plant, why can't I return to engineering?"

Gable: "Program management is a world of its own. It has its own formal and informal organizational ties. Program managers are outsiders. You'll find out. You might not be able to keep the strong personal ties you now have with your fellow employees. You'll have to force even your best friends to comply with your standards. Program managers can go from program to program, but functional departments remain intact.

"I'm telling you all this for a reason. We've worked well together the past several years. But if I sign the release so that you can work for Elliot Grey in program management, you'll be on your own, like hiring into a new company. I've already signed the release. You still have some time to think about it."

Anderson: "One thing I don't understand. With all of the good program managers we have here, why am I being given this opportunity?"

Gable: "Almost all of our program managers are over 45 years old. This resulted from our massive layoffs several years ago when we were forced to lay off the younger, inexperienced program managers. You were selected because of your age and because all of our other program managers have worked only on production-type programs. We need someone at the reins who knows R&D. Your counterpart at Lord Industries will be an R&D type. You have to fight fire with fire.

"I have an ulterior reason for wanting you to accept this position. Because of the division of authority between program management and project engineering, I need someone in program management whom I can communicate with concerning R&D work. The program managers we have now are interested only in time and cost. We need a manager who will bend over backward to get performance also. I think you're that man. You know the commitment we made to Lord when we submitted that proposal. You have to try to achieve that. Remember, this program is my baby. You'll get all the support you need. I'm tied up on another project now. But when it's over, I'll be following your work like a hawk. We'll have to get together occasionally and discuss new techniques.

"Take a day or two to think it over. If you want the position, make an appointment to see Elliot Grey, the director of program management. He'll give you the same speech I did. I'll assign Paul Evans to you as chief project engineer. He's a seasoned veteran and you should have no trouble working with him. He'll give you good advice. He's a good man."

■ The Work Begins

Gary Anderson accepted the new challenge. His first major hurdle occurred in staffing the project. The top priority given to him to bid the program did not follow through for staffing. The survival of Parks Corporation depended on the profits received from its production programs. Gary found that, in keeping with this philosophy, engineering managers (even his former boss) were reluctant to give up their key people to the Blue Spider Program. However, with a little support from Henry Gable, Gary formed an adequate staff for the program.

Right from the start, Gary was worried that the test matrix called out in the technical volume of the proposal would not produce results that could satisfy specifications. Gary had 90 days after go-ahead during which to identify the raw materials that could satisfy specification requirements. He and Paul Evans held a meeting to map out their strategy for the first few months.

Gary Anderson: "Well, Paul, we're starting out with our backs against the wall on this one. Any recommendations?"

Paul Evans: "I also have my doubts about the validity of this test matrix. Fortunately, I've been through this before. Gable thinks this is his project, and he'll sure as hell try to manipulate us. I have to report to him every morning at 7:30 a.m. with the raw data results of the previous day's testing. He wants to see it before you do. He also stated that he wants to meet with me alone.

"Lord will be the big problem. If the test matrix proves to be a failure, we're going to have to change the scope of effort. Remember, this is a firm-fixed-price contract. If we change the scope of work and do additional work in the earlier phases of the program, then we should prepare a trade-off analysis to see what we can delete downstream so as to not overrun the budget."

Anderson: "I'm going to let the other project office personnel handle the administrating work. You and I are going to live in the research labs until we get some results. We'll let the other project office personnel run the weekly team meetings."

For the next three weeks Gary and Paul spent virtually 12 hours per day, seven days a week, in the R&D lab. None of the results showed any promise. Gary kept trying to set up a meeting with Henry Gable but always found him unavailable.

During the fourth week, Gary, Paul, and the key functional department managers met to develop an alternate test matrix. The new test matrix looked good. Gary and his team worked frantically to develop a new workable schedule that would not have impact on the second milestone, which was to occur at the end of 180 days. The second milestone was the final acceptance of the raw materials and preparation of production runs of the raw materials to verify that there would be no scale-up differences between lab development and full-scale production.

Gary personally prepared all of the technical handouts for the interchange meeting. After all, he would be the one presenting the data. The technical interchange meeting was scheduled for two days. On the first day, Gary presented all of the data, including test results and the new test matrix. Lord Industries, the customer, appeared displeased with the progress to date and decided to have its own in-house caucus that evening to go over the material that was presented.

The following morning, a spokesman for Lord Industries stated its position:

> First of all, Gary, we're quite pleased to have a project manager who has such a command of technology. That's good. But every time we've tried to contact you last month, you were unavailable or had to be paged in the research laboratories. You did an acceptable job presenting the technical data, but the administrative data was presented by your project office personnel. We at Lord do not think that you're

maintaining the proper balance between your technical and administrative responsibilities. We prefer that you personally give the administrative data and your chief project engineer present the technical data.

We did not receive any agenda. Our people like to know what will be discussed, and when. We also want a copy of all handouts to be presented at least three days in advance. We need time to scrutinize the data. You can't expect us to walk in here blind and make decisions after seeing the data for ten minutes.

To be frank, we feel that the data to date is totally unacceptable. If the data does not improve, we will have no choice but to issue a work stoppage order and look for a new contractor. The new test matrix looks good, especially since this is a firm-fixed-price contract. Your company will bear the burden of all costs for the additional work. A trade-off with later work may be possible, but this will depend on the results presented at the second design review meeting, 90 days from now.

We have decided to establish a customer office at Parks to follow your work more closely. Our people feel that monthly meetings are insufficient during R&D activities. We would like our customer representative to have daily verbal meetings with you or your staff. He will then keep us posted. Obviously, we had expected to review much more experimental data than you have given us.

Many of our top-quality engineers would like to talk directly to your engineering community, without having to continually waste time by having to go through the project office. We must insist on this last point. Remember, your effort may be only $2.2 million, but our total package is $100 million. We have a lot more at stake than you people do. Our engineers do not like to get information that has been filtered by the project office. They want to help you.

And last, don't forget that you people have a contractual requirement to prepare complete minutes for all interchange meetings. Send us the original for signature before going to publication.

Although Gary was unhappy with the first team meeting, especially with the requests made by Lord Industries, he felt that the client had sufficient justification for its comments. Following the team meeting, Gary personally prepared the complete minutes. "This is absurd," thought Gary. "I've wasted almost one entire week doing nothing more than administrative paperwork. Why do we need such detailed minutes? Can't a rough summary suffice? Why is it that customers want everything documented? That's like an indication of fear. We've been completely cooperative with them. There has been no hostility between us. If we've gotten this much paperwork to do now, I hate to imagine what it will be like if we get into trouble."

■ A New Role

Gary completed and distributed the minutes to the customer and to all key team members.

For the next five weeks testing went according to plan, or at least Gary thought that it had. The results were still poor. Gary was so caught up in administrative paperwork that he hadn't found time to visit the research labs in over a month. On a Wednesday morning, Gary entered the lab to observe the morning testing. Upon arriving in the lab, Gary found Paul Evans, Henry Gable, and two technicians testing a new material, JXB-3.

Henry Gable: "Gary, your problems will soon be over. This new material, JXB-3, will permit you to satisfy specification requirements. Paul and I have been testing it for two weeks. We wanted to let you know but were afraid that if the word leaked out to the customer that we were spending their money for testing materials that were not called out in the program plan, they would probably go crazy and might cancel the contract. Look at these results. They're super!"

Gary Anderson: "Am I supposed to be the one to tell the customer now? This could cause a big wave."

Gable: "There won't be any wave. Just tell them that we did it with our own internal R&D funds. That'll please them because they'll think we're spending our own money to support their program."

Before presenting the information to Lord, Gary called a team meeting to present the new data to the project personnel. At the team meeting, one functional manager spoke out, saying, "This is a hell of a way to run a program. I like to be kept informed about everything that's happening here at Parks. How can the project office expect to get support out of the functional departments if we're kept in the dark until the very last minute? My people have been working with the existing materials for the last two months and you're telling us that it was all for nothing. Now you're giving us a material that's so new we have no information on it whatsoever. We're now going to have to play catch-up, and that's going to cost you plenty."

One week before the 180-day milestone meeting, Gary Anderson submitted the handout package to Lord Industries for preliminary review. An hour later the phone rang.

Lord Industries: We've just read your handout. Where did this new material come from? How come we were not informed that this work was going on? You know, of course, that our customer, the Army, will be at this meeting. How can we explain this to them? We're postponing the review meeting until all of our people have analyzed the data and are prepared to make a decision.

"The purpose of a review or interchange meeting is to exchange information when *both* parties have familiarity with the topic. Normally, we require almost weekly interchange meetings with our other customers because we don't trust them. We disregard this policy with Parks Corporation based on past working relationships. But with the new state of developments, you have forced us to revert to our previous position, since we now question Parks Corporation's integrity in communicating with us. At first we believed this was due to an inexperienced program manager. Now we're not sure."

Gary Anderson: "I wonder if the real reason we have these interchange meetings isn't to show our people that Lord Industries doesn't trust us. You're creating a hell of a lot of work for us, you know."

Lord Industries: "You people put yourself in this position. Now you have to live with it."

Two weeks later, Lord reluctantly agreed that the new material offered the greatest promise. Three weeks later, the design review meeting was held. The Army was definitely not pleased with the prime contractor's recommendation to put a new, untested material into a multimillion-dollar effort.

■ The Communications Breakdown

During the week following the design review meeting, Gary planned to make the first verification mix in order to establish final specifications for selection of the raw materials. Unfortunately, the manufacturing plans were a week behind schedule, primarily because of Gary, since he had decided to reduce costs by accepting the responsibility for developing the bill of materials himself.

Gary Anderson called a meeting to consider rescheduling of the mix.

Gary Anderson: "As you know, we're about a week to 10 days behind schedule. We'll have to reschedule the verification mix for late next week."

Production manager: "Our resources are committed until a month from now. You can't expect to simply call a meeting and have everything reshuffled for the Blue Spider Program. We should have been notified earlier. Engineering has the responsibility for preparing the bill of materials. Why aren't they ready?"

Engineering integration: "We were never asked to prepare the bill of materials. But I'm sure that we could get it out if we work our people overtime for the next two days."

Anderson: "When can we remake the mix?"

Production manager: "We have to redo at least 500 sheets of paper every time we reschedule mixes. Not only that, we have to reschedule people on all three shifts. If we are to reschedule your mix, it will have to be performed on overtime. That's going to increase your costs. If that's agreeable to you, we'll try it. But this will be the first and last time that production will bail you out. There are procedures that have to be followed."

Testing engineer: "I've been coming to these meetings since we kicked off this program. I think I speak for the entire engineering division when I say that the role the director of engineering is playing in this program is suppressing individuality among our highly competent personnel. In new projects, especially those involving R&D, our people are not apt to stick their necks out. Now our people are becoming ostriches. If they're impeded from contributing, even in their own slight way, then you'll probably lose them before the project gets completed. Right now I feel that I'm wasting my time here. All I need are minutes of the team meetings and I'll be happy. Then I won't have to come to these pretend meetings anymore."

The purpose of the verification mix was to make a full-scale production run of the material to verify that there would be no material property changes in scale-up from the small mixes made in the R&D laboratories. After testing, it became obvious that the wrong lots of raw materials were used in the production verification mix.

Lord Industries called a meeting for an explanation of why the mistake had occurred and what the alternatives were.

Lord Industries: "Why did the problem occur?"

Gary Anderson: "Well, we had a problem with the bill of materials. The result was that the mix had to be made on overtime. And when you work people on overtime, you have to be willing to accept mistakes as being a way of life. The energy cycles of our people are slow during the overtime hours."

Lord Industries: "The ultimate responsibility has to be with you, the program manager. We at Lord think that you're spending too much time doing and not enough time managing. As the prime contractor, we have a hell of a lot more at stake than you do. From now on we want documented weekly technical interchange meetings and closer interaction by our quality control section with yours."

Anderson: "These additional team meetings are going to tie up our key people. I can't spare people to prepare handouts for weekly meetings with your people."

Lord Industries: "Team meetings are a management responsibility. If Parks does not want the Blue Spider Program, I'm sure we can find another subcontractor. All you, Gary, have to do is give up taking the material vendors to lunch, and you'll have plenty of time for handout preparation."

Gary left the meeting feeling as if he had gotten raked over the coals. For the next two months, Gary worked 16 hours a day, almost every day. Gary did not want to burden his staff with the responsibility of the handouts, so he began preparing them himself. He could have hired additional staff, but with such a tight budget, and having to remake the verification mix, cost overruns appeared inevitable.

As the end of the seventh month approached, Gary was feeling pressure from within Parks Corporation. The decision-making process appeared to be slowing down, and Gary found it more and more difficult to motivate his people. In fact, the grapevine was referring to the Blue Spider Project as a loser, and some of his key people acted as if they were on a sinking ship.

By the time the eighth month rolled around, the budget had nearly been expended. Gary was tired of doing everything himself. "Perhaps I should have stayed an engineer," he thought. He and Elliot Grey had a meeting to see what could be salvaged. Grey agreed to get Gary additional corporate funding to complete the project. "But performance must be met, since there is a lot riding on the Blue Spider Project," asserted Grey. Gary called a team meeting to identify the program status.

Gary Anderson: "It's time to map out our strategy for the remainder of the program. Can engineering and production adhere to the schedule I have laid out before you?"

Team member, engineering: "This is the first time I've seen this schedule. You can't expect me to make a decision in the next 10 minutes and commit the resources of my department. We're getting a little unhappy being kept in the dark until the last minute. What happened to effective planning?"

Anderson: "We still have effective planning. We must adhere to the original schedule, or at least try to adhere to it. This revised schedule will do that."

Team member, engineering: "Look, Gary! When a project gets in trouble, it is usually the functional departments that come to the rescue. But if we're kept in the dark, then how can you expect us to come to your rescue? My boss wants to know, well in advance, every decision that you're contemplating with regard to our departmental resources. Right now, we—"

Anderson: "Granted, we may have had a communications problem. But now we're in trouble and have to unite forces. What is your impression as to whether your department can meet the new schedule?"

Team member, engineering: "When the Blue Spider Program first got in trouble, my boss exercised his authority to make all departmental decisions regarding the program himself. I'm just a puppet. I have to check with him on everything."

Team member, production: "I'm in the same boat, Gary. You know we're not happy having to reschedule our facilities and people. We went through this once before. I also have to check with my boss before giving you an answer about the new schedule."

The following week, the verification mix was made. Testing proceeded according to the revised schedule, and it looked as if the total schedule milestones could be met, provided that specifications could be adhered to.

Because of the revised schedule, some of the testing had to be performed on holidays. Gary wasn't pleased with asking people to work on Sundays and holidays, but he had no choice, since the test matrix called for testing to be accomplished at specific times after end of mix.

A team meeting was called on Wednesday to resolve the problem of who would work on the holiday, which would occur on Friday, as well as staffing Saturday and Sunday. During the team meeting, Gary became quite disappointed. Phil Rodgers, who had been Gary's test engineer since the project started, was assigned to a new project that the grapevine called Gable's new adventure. His replacement was a relatively new man, with the company only eight months. For an hour and a half, the team members argued about little problems and continually avoided the major question, stating that they would first have to coordinate commitments with their bosses. It was obvious to Gary that his team members were afraid to make major decisions and therefore ate up a lot of time on trivial problems.

On the following day, Thursday, Gary went to see the department manager responsible for testing, in hopes that he could use Phil Rodgers that weekend.

Department manager: "I have specific instructions from the boss [director of engineering] to use Phil Rodgers on the new project. You'll have to see the boss if you want him back."

Gary Anderson: "But we have testing that must be accomplished this weekend. Where's the new man you assigned yesterday?"

Department manager: "Nobody told me you had testing scheduled for this weekend. Half of my department is already on an extended weekend vacation, including Phil Rodgers and the new man. How come I'm always the last to know when we have a problem?"

Anderson: "The customer is flying down his best people to observe this weekend's tests. It's too late to change anything. You and I can do the testing."

Department manager: "Not on your life. I'm staying as far away as possible from the Blue Spider Project. I'll get you someone, but it won't be me. That's for sure!"

The weekend's testing went according to schedule. The raw data was made available to the customer under the stipulation that the final company position would be announced at the end of the next month, after the functional departments had a chance to analyze it.

Final testing was completed during the second week of the ninth month. The initial results looked excellent. The materials were within contract specifications, and although they were new, both Gary and Lord's management felt that there would be little difficulty in convincing the Army that this was the way to go. Henry Gable visited Gary and congratulated him on a job well done.

All that now remained was the making of four additional full-scale verification mixes in order to determine how much deviation there would be in material properties between full-size production-run mixes. Gary tried to get the customer to concur (as part of the original trade-off analysis) that two of the four production runs could be deleted. Lord Industries' management refused, insisting that contractual requirements must be met at the expense of the contractor.

The following week, Elliot Grey called Gary in for an emergency meeting concerning expenditures to date.

Elliot Grey: "Gary, I just received a copy of the financial planning report for last quarter in which you stated that both the cost and performance of the Blue Spider Project were 75 percent complete. I don't think you realize what you've done. The target profit on the program was $200,000. Your memo authorized the vice president and general manager to book 75 percent of that, or $150,000, for corporate profit spending for stockholders. I was planning on using all $200,000 together with the additional $300,000 I personally requested from corporate headquarters to bail you out. Now I have to go back to the vice president and general manager and tell them that we've made a mistake and that we'll need an additional $150,000."

Gary Anderson: "Perhaps I should go with you and explain my error. Obviously, I take all responsibility."

Grey: "No, Gary. It's our error, not yours. I really don't think you want to be around the general manager when he sees red at the bottom of the page. It takes an act of God to get money back once corporate books it as profit. Perhaps you should reconsider project engineering as a career instead of program management. Your performance hasn't exactly been sparkling, you know."

Gary returned to his office quite disappointed. No matter how hard he worked, the bureaucratic red tape of project management seemed always to do him in. But late that afternoon, Gary's disposition improved. Lord Industries called to say that, after consultation with the Army, Parks Corporation would be awarded a sole-source contract for qualification and production of Spartan missile components using the new longer-life raw materials. Both Lord and the Army felt that the sole-source contract was justified, provided that continued testing showed the same results, since Parks Corporation had all of the technical experience with the new materials.

Gary received a letter of congratulations from corporate headquarters but no additional pay increase. The grapevine said that a substantial bonus was given to the director of engineering.

During the 10th month, results were coming back from the accelerated aging tests performed on the new materials. The results indicated that although the new materials would meet specifications, the age life would probably be less than five years. These numbers came as a shock to Gary. He and Paul Evans had a conference to determine the best strategy to follow.

Gary Anderson: "Well, I guess we're now in the fire instead of the frying pan. Obviously, we can't tell Lord Industries about these tests. We ran them on our own. Could the results be wrong?"

Paul Evans: "Sure, but I doubt it. There's always margin for error when you perform accelerated aging tests on new materials. There can be reactions taking place that we know nothing about. Furthermore, the accelerated aging tests may not even correlate well with actual aging. We must form a company position on this as soon as possible."

Anderson: "I'm not going to tell anyone about this, especially Henry Gable. You and I will handle this. It will be my throat if word of this leaks out. Let's wait until we have the production contract in hand."

Evans: "That's dangerous. This has to be a company position, not a project office position. We had better let them know upstairs."

Anderson: "I can't do that. I'll take all responsibility. Are you with me on this?"

Evans: "I'll go along. I'm sure I can find employment elsewhere when we open Pandora's box. You had better tell the department managers to be quiet also."

Two weeks later, as the program was winding down into the testing for the final verification mix and final report development, Gary received an urgent phone call asking him to report immediately to Henry Gable's office.

Henry Gable: "When this project is over, you're through. You'll never hack it as a program manager or possibly even a good project engineer. We can't run projects around here without honesty and open communications. How the hell do you expect top management to support you when you start censoring bad news to the top? I don't like surprises. I like to get the bad news from the program manager and project engineers, not secondhand from the customer. And, of course, we cannot forget the cost overrun. Why didn't you take some precautionary measures?"

Gary Anderson: "How could I when you were asking our people to do work such as accelerated aging tests that would be charged to my project and was not part of program plan? I don't think I'm totally to blame for what's happened."

Gable: "Gary, I don't think it's necessary to argue the point any further. I'm willing to give you back your old job, in engineering. I hope you didn't lose too many friends while working in program management. Finish up final testing and the program report. Then I'll reassign you."

Gary returned to his office and put his feet up on the desk. "Well," he thought, "perhaps I'm better off in engineering. At least I can see my wife and kids once in a while." As he began writing the final report, the phone rang.

Functional manager: "Hello, Gary. I just thought I'd call to find out what charge number you want us to use for experimenting with this new procedure to determine accelerated age life."

Gary Anderson: "Don't call me! Call Gable. After all, the Blue Spider Project is his baby."

■ Questions

1. Can assessments determine if the project managers have the right qualifications?

2. Can a PMMM assessment determine the maturity of a firm's competitive bidding processes?

3. Can an assessment determine if a PM should head up a proposal preparation team?

4. Can an assessment determine an organization's ability to maintain truth and honesty during competitive bidding?

5. Can a PMMM be used for assessing both project management and program management?

6. Can a PMMM assess if projects are properly staffed?

7. Can a PMMM determine if the project management is maintaining the right mix between technical and administrative responsibilities?

8. How should a company respond to an assessment that reports a low level of maturity when the company believes it was the result of changes that had to be made because the customers did not trust them?

9. Can a PMMM assessment determine the maturity of customer involvement in projects?

10. Should an assessment determine if the project manager has the correct amount of authority?

11. Should maturity assessment evaluate to whom the project manager reports or should report to?

12. Should an assessment determine if the project manager is spending money correctly?

► Case 8: Corwin Corporation

By June 2003, Corwin Corporation had grown into a $950 million-per-year corporation with an international reputation for manufacturing low-cost, high-quality rubber components. Corwin maintained more than a dozen different product lines, all of which were sold as off-the-shelf items in department stores, hardware stores, and automotive parts distributors. The name "Corwin" was now synonymous with "quality." This provided management with the luxury of having products that had extremely long life cycles.

Organizationally, Corwin had maintained the same structure for more than 15 years (see Figure 14.1). The top management of Corwin Corporation was highly conservative and believed in using a marketing approach to find new markets for existing product lines rather than exploring for new products. Under this philosophy, Corwin maintained a small R&D group whose mission was simply to evaluate state-of-the-art technology and its application to existing product lines.

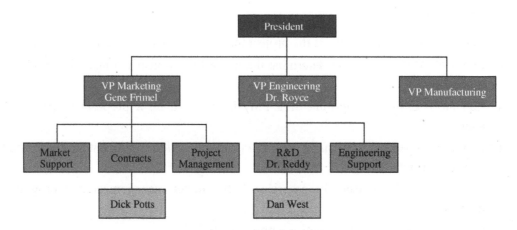

Figure 14.1 Organizational chart for Corwin Corporation.

Corwin's reputation was so good that it continually received inquiries about the manufacturing of specialty products. Unfortunately, the conservative nature of Corwin's management created a "don't rock the boat" atmosphere opposed to taking any type of risks. A management policy was established to evaluate all specialty-product requests. The policy required answering yes to the following questions:

- Will the specialty product provide the same profit margin (20%) as existing product lines?

- Is there a chance for follow-on contracts?

- Can the specialty product be developed into a product line?

- Can the specialty product be produced with minimum disruption to existing product lines and manufacturing operations?

These stringent requirements forced Corwin not to bid on more than 90% of all specialty-product inquiries.

Corwin Corporation was a marketing-driven organization, although manufacturing often had different ideas. Almost all decisions were made by marketing, with the exception of product pricing and estimating, which was a joint undertaking between manufacturing and marketing. Engineering was considered as merely a support group to marketing and manufacturing.

For specialty products, the project managers always came out of marketing, even during the R&D phase of development. The company's approach was that if the specialty product should mature into a full product line, then there should be a product line manager assigned right at the onset.

■ The Peters Company Project

In 2000, Corwin accepted a specialty-product assignment from Peters Company because of the potential for follow-on work. In 2001, 2002, and again in 2003, profitable follow-on contracts were received, and a good working relationship developed, despite Peters' reputation for being a difficult customer to work with.

On December 7, 2002, Gene Frimel, the vice president of marketing at Corwin, received an unusual phone call from Dr. Frank Delia, the marketing vice president at Peters Company.

Frank Delia: "Gene, I have a rather strange problem on my hands. Our R&D group has $250,000 committed for research toward development of a new rubber product material, and we simply do not have the available personnel or talent to undertake the project. We have to go outside. We'd like your company to do the work. Our testing and R&D facilities are already overburdened."

Gene Frimel: "Well, as you know, Frank, we are not a research group, even though we've done this once before for you. And furthermore, I would never be able to sell our management on such an undertaking. Let some other company do the R&D work and then we'll take over on the production end."

Delia: "Let me explain our position on this. We've been burned several times in the past. Projects like this generate several patents, and the R&D company almost always requires that our contracts give it royalties or first refusal for manufacturing rights."

Frimel: "I understand your problem, but it's not within our capabilities. This project, if undertaken, could disrupt parts of our organization. We're already operating lean in engineering."

Delia: "Look, Gene! The bottom line is this: We have complete confidence in your manufacturing ability to such a point that we're willing to commit to a five-year production contract if the product can be developed. That makes it extremely profitable for you."

Frimel: "You've just gotten me interested. What additional details can you give me?"

Delia: "All I can give you is a rough set of performance specifications that we'd like to meet. Obviously, some trade-offs are possible."

Frimel: "When can you get the specification sheet to me?"

Delia: "You'll have it tomorrow morning. I'll ship it overnight express."

Frimel: "Good! I'll have my people look at it, but we won't be able to get you an answer until after the first of the year. As you know, our plant is closed down for the last two weeks in December, and most of our people have already left for extended vacations."

Delia: "That's not acceptable! My management wants a signed, sealed, and delivered contract by the end of this month. If this is not done, corporate will reduce our budget for 2003 by $250,000, thinking that we've bitten off more than we can chew. Actually, I need your answer within 48 hours so that I'll have some time to find another source."

Frimel: "You know, Frank, today is December 7, Pearl Harbor Day. Why do I feel as though the sky is about to fall in?"

Delia: "Don't worry, Gene! I'm not going to drop any bombs on you. Just remember, all that we have available is $250,000, and the contract must be a firm-fixed-price effort. We anticipate a six-month project with $125,000 paid on contract signing and the balance at project termination."

Frimel: "I still have that ominous feeling, but I'll talk to my people. You'll hear from us with a go or no-go decision within 48 hours. I'm scheduled to go on a Caribbean cruise, and my wife and I are leaving this evening. One of my people will get back to you on this matter."

Gene Frimel had a problem. All bid and no-bid decisions were made by a four-person committee composed of the president and the three vice presidents. The president and the vice president for manufacturing were on vacation. Frimel met with Dr. Lindsay Royce, the vice president of engineering, and explained the situation.

Lindsay Royce: "You know, Gene, I totally support projects like this because it would help our technical people grow intellectually. Unfortunately, my vote never appears to carry any weight."

Gene Frimel: "The profitability potential as well as the development of good customer relations makes this attractive, but I'm not sure we want to accept such a risk. A failure could easily destroy our good working relationship with Peters Company."

Royce: "I'd have to look at the specification sheets before assessing the risks, but I would like to give it a shot."

Frimel: "I'll try to reach our president by phone."

By late afternoon, Frimel was fortunate enough to be able to contact the president and received a reluctant authorization to proceed. The problem now was how to prepare a proposal within the next two or three days and be ready to make an oral presentation to Peters Company.

Gene Frimel: "The boss gave his blessing, Royce, and the ball is in your hands. I'm leaving for vacation, and you'll have total responsibility for the proposal and presentation. Delia wants the presentation this weekend. You should have his specification sheets tomorrow morning."

Lindsay Royce: "Our R&D director, Dr. Reddy, left for vacation this morning. I wish he were here to help me price out the work and select the project manager. I assume that, in this case, the project manager will come out of engineering rather than marketing."

Frimel: "Yes, I agree. Marketing should not have any role in this effort. It's your baby all the way. And as for the pricing effort, you know our bid will be for $250,000. Just work backward to justify the numbers. I'll assign one of our contracting people to assist you in the pricing. I hope I can find someone who has experience in this type of effort. I'll call Delia and tell him we'll bid it with an unsolicited proposal."

Royce selected Dan West, one of the R&D scientists, to act as the project leader. Royce had severe reservations about doing this without the R&D director, Dr. Reddy, being actively involved. But with Reddy on vacation, Royce had to make an immediate decision.

On the following morning, the specification sheets arrived and Royce, West, and Dick Potts, a contracts specialist, began preparing the proposal. West prepared the direct labor hours, and Royce provided the costing data and pricing rates. Potts, being completely unfamiliar with this type of effort, simply acted as an observer and provided legal advice when necessary. Potts allowed Royce to make all decisions, even though Potts was considered the president's official representative.

Finally completed two days later, the proposal was actually a 10-page letter that simply contained the cost summaries (see Table 14.1) and the engineering intent. West estimated that 30 tests would be required. The test matrix described the test conditions only for the first five tests. The remaining 25 test conditions would be determined at a later date, jointly by Peters and Corwin personnel.

Table 14.1 Proposal cost summaries.

Direct labor and support	$ 30,000
Testing (30 tests at $2,000 each)	60,000
Overhead at 100%	90,000
Materials	30,000
General and administrative (G&A), 10%	21,000
Total	$231,000
Profit	19,000
Total	$250,000

On Sunday morning, a meeting was held at Peters Company, and the proposal was accepted. Delia gave Royce a letter of intent authorizing Corwin Corporation to begin working on the project immediately. The final contract would not be available for signing until late January, and the letter of intent simply stated that Peters Company would assume all costs until such time that the contract was signed or the effort terminated.

West was truly excited about being selected as the project manager and being able to interface with the customer, a luxury that was usually given only to marketing personnel. Although Corwin Corporation was closed for two weeks over Christmas, West still went into the office to prepare the project schedules and to identify the support he would need in the other areas, thinking that if he presented this information to management on the first day back to work, they would be convinced that he had everything under control.

■ The Work Begins

On the first working day in January 2003, a meeting was held with the three vice presidents and Dr. Reddy to discuss the support needed for the project. (West was not in attendance at this meeting, although all participants had a copy of his memo.)

Dr. Reddy: "I think we're heading for trouble in accepting this project. I've worked with Peters Company previously on R&D efforts, and they're tough to get along with. West is a good man, but I would never have assigned him as the project leader. His expertise is in managing internal rather than external projects. But, no matter what happens, I'll support West the best I can."

Lindsay Royce: "You're too pessimistic. You have good people in your group and I'm sure you'll be able to give him the support he needs. I'll try to look in on the project every so often. West will still be reporting to you for this project. Try not to burden him too much with other work. This project is important to the company."

West spent the first few days after vacation soliciting the support that he needed from the other line groups. Many of the other groups were upset that they had not been informed earlier and were unsure as to what support they could provide. West met with Reddy to discuss the final schedules. Reddy said:

> Your schedules look pretty good, Dan. I think you have a good grasp on the problem. You won't need very much help from me. I have a lot of work to do on other activities, so I'm just going to be in the background on this project. Drop me a note every once in a while telling me what's going on. I don't need anything formal. Just a paragraph or two will suffice.

By the end of the third week, all of the raw materials had been purchased, and initial formulations and testing were ready to begin. In addition, the contract was ready for signature. The contract contained a clause specifying that Peters Company had the right to send an in-house representative into Corwin Corporation for the duration of the project. Peters Company informed Corwin that Patrick Ray would be the in-house representative, reporting to Delia, and would assume his responsibilities on or about February 15.

By the time Pat Ray appeared at Corwin Corporation, West had completed the first three tests. The results were not what was expected but indicated that Corwin was heading in the right direction. Pat Ray's interpretation of the tests was completely opposite West's. Ray thought that Corwin was "way off base" and that redirection was needed.

Pat Ray: "Look, Dan! We have only six months to do this effort, and we shouldn't waste our time on marginally acceptable data. These are the next five tests I'd like to see performed."

Dan West: "Let me look over your request and review it with my people. That will take a couple of days, and meanwhile, I'm going to run the other two tests as planned."

Ray's arrogant attitude bothered West. However, West decided that the project was too important to knock heads with Ray and decided to cater to Ray the best he could. This was not exactly the working relationship that West expected to have with the in-house representative.

West reviewed the test data and the new test matrix with engineering personnel, who felt that the test data was inconclusive as yet and preferred to withhold their opinion until the results of the fourth and fifth tests were made available. Although this displeased Ray, he agreed to wait a few more days if it meant getting Corwin Corporation on the right track.

The fourth and fifth tests appeared to be marginally acceptable, just as the first three had been. Corwin's engineering people analyzed the data and made their recommendations.

Dan West: "Pat, my people feel that we're going in the right direction and that our path has greater promise than your test matrix."

Pat Ray: "As long as we're paying the bills, we're going to have a say in what tests are conducted. Your proposal stated that we would work together in developing the other test conditions. Let's go with my test matrix. I've already reported back to my boss that the first five tests were failures and that we're changing the direction of the project."

West: "I've already purchased $30,000 worth of raw materials. Your matrix uses other materials and will require additional expenditures of $12,000."

Ray: "That's your problem. Perhaps you shouldn't have purchased all of the raw materials until we agreed on the complete test matrix."

During the month of February, West conducted 15 tests, all under Ray's direction. The tests were scattered over such a wide range that no valid conclusions could be drawn. Ray continued sending reports back to Delia confirming that Corwin was not producing beneficial results and that there was no indication that the situation would reverse itself. Delia ordered Ray to take any steps necessary to ensure a successful completion of the project.

Ray and West met again as they had done for each of the past 45 days to discuss the status and direction of the project.

Pat Ray: "Dan, my boss is putting tremendous pressure on me for results, and thus far I've given him nothing. I'm up for promotion in a couple of months and I can't let this project stand in my way. It's time to completely redirect the project."

Dan West: "Your redirection of the activities is playing havoc with my scheduling. I have people in other departments who just cannot commit to this continual rescheduling. They blame me for not communicating with them when, in fact, I'm embarrassed to."

Ray: "Everybody has their problems. We'll get this problem solved. I spent this morning working with some of your lab people in designing the next 15 tests. Here are the test conditions."

West: "I certainly would have liked to be involved with this. After all, I thought I was the project manager. Shouldn't I have been at the meeting?"

Ray: "Look, Dan! I really like you, but I'm not sure that you can handle this project. We need some good results immediately, or my neck will be stuck out for the next four months. I don't want that. Just have your lab personnel start on these tests, and we'll get along fine. Also, I'm planning on spending a great deal of time in your lab area. I want to observe the testing personally and talk to your lab personnel."

West: "We've already conducted 20 tests, and you're scheduling another 15 tests. I priced out only 30 tests in the proposal. We're heading for a cost overrun condition."

Ray: "Our contract is a firm-fixed-price effort. Therefore, the cost overrun is your problem."

West met with Dr. Reddy to discuss the new direction of the project and potential cost overruns. West brought along a memo projecting the costs through the end of the third month of the project. (See Table 14.2)

Table 14.2 Projected cost summary at the end of the third month.

	Original proposal cost summary for six-month project	Total project costs projected at end of third month
Direct labor/support	$ 30,000	$ 15,000
Testing	60,000 (30 tests)	70,000 (35 tests)
Overhead	90,000 (100%)	92,000 (120%)*
Materials	30,000	50,000
G&A	21,000 (10%)	22,700 (10%)
Totals	$231,000	$249,700

*Total engineering overhead was estimated at 100%, whereas the R&D overhead was 120%.

Reddy told West, "I'm already overburdened on other projects and won't be able to help you out. Royce picked you to be the project manager because he felt that you could do the job. Now, don't let him down. Send me a brief memo next month explaining the situation, and I'll see what I can do. Perhaps the situation will correct itself."

During March, the third month of the project, West received almost daily phone calls from the people in the lab stating that Pat Ray was interfering with their job. In fact, one phone call stated that Ray had changed the test conditions from what was agreed on in the latest test matrix. When West confronted Ray on his meddling, Ray asserted that Corwin personnel were very unprofessional in their attitude and that he thought this was being carried down to the testing as well. Furthermore, Ray demanded that one of the functional employees be removed immediately from the project because of incompetence. West stated that he would talk to the employee's department manager. Ray, however, felt that this would be useless and said, "Remove him or else!" The functional employee was removed from the project.

By the end of the third month, most Corwin employees were disenchanted with the project and were looking for other assignments. West attributed this to Ray's harassment of the employees. To aggravate the situation even further, Ray met with Royce and Reddy and demanded that West be removed and a new project manager be assigned.

Royce refused to remove West as project manager and ordered Reddy to take charge and help West get the project back on track. Reddy said: "You've kept me in the dark concerning this project, West. If you want me to help you, as Royce requested, I'll need all the information tomorrow, especially the cost data. I'll expect you in my office tomorrow morning at 8:00 a.m. I'll bail you out of this mess."

West prepared the projected cost data for the remainder of the work and presented the results to Dr. Reddy. (See Table 14.3) Both West and Reddy agreed that the project was now out of control and severe measures would be required to correct the situation, in addition to more than $250,000 in corporate funding.

Table 14.3 Estimate of total project-completion costs.

Direct labor/support	$ 47,000*
Testing (60 tests)	120,000
Overhead (120%)	200,000
Materials	103,000
G&A	47,000
	$517,000
Peters contract	250,000
Overrun	$267,000

*Includes Dr. Reddy.

Dr. Reddy: "Dan, I've called a meeting for 10:00 a.m. with several of our R&D people to completely construct a new test matrix. This is what we should have done right from the start."

Dan West: "Shouldn't we invite Ray to attend this meeting? I'm sure he'd want to be involved in designing the new test matrix."

Reddy: "I'm running this show now, not Ray! Tell Ray that I'm instituting new policies and procedures for in-house representatives. He's no longer authorized to visit the labs at his own discretion. He must be accompanied by either you or me. If he doesn't like these rules, he can get out. I'm not going to allow that guy to disrupt our organization. We're spending our money now, not his."

West met with Ray and informed him of the new test matrix as well as the new policies and procedures for in-house representatives. Ray was furious over the new turn of events and stated that he was returning to Peters Company for a meeting with Delia.

On the following Monday, Frimel received a letter from Delia stating that Peters Company was officially canceling the contract. The reasons given by Delia were as follows:

1. Corwin had produced absolutely no data that looked promising.
2. Corwin continually changed the direction of the project and did not appear to have a systematic plan of attack.
3. Corwin did not provide a project manager capable of handling such a project.
4. Corwin did not provide sufficient support for the in-house representative.
5. Corwin's top management did not appear to be sincerely interested in the project and did not provide sufficient executive-level support.

Royce and Frimel met to decide on a course of action in order to sustain good working relations with Peters Company. Frimel wrote a strong letter refuting all of the accusations in the Peters letter, but to no avail. Even the fact that Corwin was willing to spend $250,000 of its own funds had no bearing on Delia's decision. The damage was done. Frimel was now thoroughly convinced that a contract should not be signed on Pearl Harbor Day.

■ Questions

1. All projects involve some degree of risk management. Can assessments measure the firm's ability to take risks and the processes used for risk management?
2. Corwin Corporation has both product lines and specialty products. Should the same PMMM assessments be used for both?
3. Assessment instruments can measure project governance and approvals. Should the same assessments be used for approvals needed during project execution and competitive bidding activities?
4. Companies that perform "backward pricing" often put pressure on project teams and force them to take unnecessary risks. Can the source of risks be measured and identified as part of assessments?
5. Some companies allow onsite representatives from their clients to reside in the building during the execution of the project. Should this be covered as part of assessments?

▶ Case 9: The Trophy Project

The ill-fated Trophy Project was in trouble right from the start. Laina Reichart, who had been an assistant project manager, was involved with the project from its conception. When the Trophy Project was accepted by the company, Reichart was assigned as the project manager. The program schedules started to slip from day 1, and expenditures were excessive. Reichart found that the functional managers were charging direct labor time to her project but working on their own pet projects. When she complained of this, she was told not to meddle in the functional managers' allocation of resources and budgeted expenditures. After approximately six months, Reichart was requested to make a progress report directly to corporate and division staffs.

Reichart took this opportunity to bare her soul. The report substantiated that the project was forecasted to be one complete year behind schedule. Reichart's staff, as supplied by the line managers, was inadequate to maintain the current pace, let alone make up any time that had already been lost. The estimated cost at completion at this interval showed a cost overrun of at least 20%. This was Reichart's first opportunity to tell her story to people who were in a position to correct the situation. The result of Reichart's frank, candid evaluation of the Trophy Project was very predictable. Nonbelievers finally saw the light, and line managers realized that they had a role to play in the completion of the project. Most of the problems were now out in the open and could be corrected with adequate staffing and resources. Corporate staff ordered immediate remedial action and staff support to provide Reichart a chance to bail out her program.

The results were not at all what Reichart had expected. She no longer reported to the project office; she now reported directly to the operations manager. Corporate staff's interest in the project became very intense, requiring a 7:00 a.m. meeting every Monday for complete review of the project status and plans for recovery. Reichart found herself spending more time preparing paperwork, reports, and projections for her Monday morning meetings than she did administering the Trophy Project. The main concern of corporate was to get the project back on schedule. Reichart spent many hours preparing the recovery plan and establishing staffing requirements to bring the program back onto the original schedule.

Group staff, in order to closely track the progress of the Trophy Project, assigned an assistant program manager. The assistant program manager determined that a sure cure for the Trophy Project would be to computerize the various problems and track the progress through a very complex computer program. Corporate provided Reichart with 12 additional staff members to work on the computer program. In the meantime, nothing changed. The functional managers still did not provide adequate staff for recovery, because they assumed that the additional resources Reichart had received from corporate would accomplish that task.

After approximately $50,000 was spent on the computer program to track the problems, it was found that the computer could not handle the program objectives. Reichart discussed this problem with a computer supplier and found that $15,000 more was required for programming and additional storage capacity. It would take two months for installation of the additional storage capacity and completion of the programming. At this point, the decision was made to abandon the computer program.

Reichart was now a year and a half into the program with no prototype units completed. The program was still nine months behind schedule, with the overrun projected at 40% of budget. The customer had been receiving reports on a timely basis and was well aware that the Trophy Project was behind schedule. Reichart had spent a great deal of time with the customer explaining the problems and the plan for recovery. Another problem that Reichart had to contend with was that the vendors who were supplying components for the project were also running behind schedule.

One Sunday morning, while Reichart was in her office putting together a report for the client, a corporate vice president came in. "Reichart," he said, "in any project I look at the top sheet of paper, and the person whose name appears at the top of the sheet is the one I hold responsible. For this project, your name appears at the top of the sheet. If you cannot bail this thing out, you are in serious trouble in this corporation." Reichart did not know which way to turn or what to say. She had no control over the functional managers who were creating the problems, but she was the person who was being held responsible.

After another three months, the customer, becoming impatient, realized that the Trophy Project was in serious trouble and requested that the division general manager and his entire staff visit the customer's plant to give a progress and get-well report within a week. The division general manager called Reichart into his office and said, "Reichart, go visit our customer. Take three or four functional line people with you and try to placate him with whatever you feel is necessary." Reichart and four functional line people visited the customer and gave a four-and-a-half-hour presentation defining the problems and the progress to that point. The customer was very polite and even commented that it was an excellent presentation, but the content was totally unacceptable. The program was still six to eight months late, and the customer demanded progress reports on a weekly basis. The customer made arrangements to assign a representative in Reichart's department to be on-site at the project on a daily basis and to interface with Reichart and her staff as required. After this turn of events, the program became very hectic.

The customer representative demanded constant updates and problem identification and then became involved in attempting to solve these problems. This involvement created many changes in the program and the product in order to eliminate some of the problems. Reichart had trouble with the customer and did not agree with the changes in the program. She expressed her disagreement vocally when, in many cases, the customer felt the changes were at no cost. This caused a deterioration of the relationship between client and producer.

One morning Reichart was called into the division general manager's office and introduced to Mr. "Red" Baron. Reichart was told to turn over the reins of the Trophy Project to Red immediately. "Reichart, you will be temporarily reassigned to some other division within the corporation. I suggest you start looking outside the company for another job." Reichart looked at Red and asked, "Who did this? Who shot me down?"

Red was program manager on the Trophy Project for approximately six months, after which, by mutual agreement, he was replaced by a third project manager. The customer reassigned his local program manager to another project. With the new team, the Trophy Project was finally completed one year behind schedule and at a 40% cost overrun.

■ Questions

1. Can a PMMM assessment measure the ability of a company to recover a failing project?

2. Can an assessment measure project staffing changes, whether for good or bad, while trying to recover a failing project?

3. Can or should an assessment measure the effectiveness of the working relationship between the project managers and the functional managers?

4. Can an assessment measure the quality of project planning activities?

5. Can an assessment measure executive-level commitment to a project?

6. Can an assessment measure the cooperation of team members during a project?

The Kerzner Project Management Maturity Model

The Kerzner Project Management Maturity Model (KPMMM) shown in this appendix represents a multinational company with many divisions. This is just an example of what the report might look like. A report can be generated for just one company or just one unit within the company. The assessment has a database of over 150 companies for benchmarking purposes. For additional information on using the KPMMM for an assessment on your firm's maturity level, contact Lori Milhaven at the International Institute for Learning (lori.milhaven@iil.com). The material in the report is copyrighted to the International Institute for Learning. Reproduced by permission.

▶ XXXX KPMMM

■ Levels 1-5 Final Assessment Report

Further to XXXX's recent KPMMM intermediate, this final report is being published to discuss all findings from XXXX's recent KPMMM assessment in granular detail.

Contents

▶ Introduction

■ Purpose of this Document

This document is a report of the results of the project management assessment performed by XXXX between MONTH to MONTH, YEAR, using IIL's Kerzner PM Maturity Model and online KPMMM™ tool set.

■ Goals and Objectives of the Assessment

The goal driving the assessment is to improve project performance in the organization and become a world-class organization.

XXXX's objective for this assessment is to obtain an objective analysis of the organization's current project management process maturity, compare scores of other organizations, and identify prescriptive actions to increase the project management maturity level throughout the organization. Note that the prescriptive actions referenced in this report are based on assessment results that are an indication of the state of the organization but must be verified by objectively assessing performance, identifying issues and doing cause and effect analysis and SWOT analysis to provide a basis for action planning. This assessment is only one part of the process.

Performance improvement is the goal. Continuous, well-managed effort is needed to achieve the right level of improvement for the right cost.

■ Report Organization

The report is organized to first provide background, a description of the assessment process, and a high-level description of the Kerzner Project Management Maturity Model (KPMM™) as a foundation for the analysis of the results.

The results are presented in an executive overview followed by comparisons, suggested actions, and a more detailed analysis.

■ About XXXX

This report represents a multinational XXXX INDUSTRY company employing over XXXX people worldwide. It is vertically integrated and operates in all areas of the XXXX industry, (BRIEF DESCRIPTION OF THE COMPANY MAIN BUSINESS UNITS/ OPERATIONS/ PROCESSES).

XXXX is continuing its program to increase the practical use of project management concepts, tools, and techniques in its projects and to establish a companywide framework that would promote consistency while allowing for flexibility across business units and diverse types of projects.

■ The Kerzner Project Management Maturity Model—Overview

The Kerzner Project Management Maturity Model is the foundation for the assessment. The model identifies five levels of maturity:

1. Common language—The foundation for improvement is based on understanding what PM is and being able to discuss and analyze it effectively. To what degree are the individuals in the organization knowledgeable about project management and to what degree can they discuss it in common terms based on the PMI *PMBOK® Guide*?

2. Common processes—How far has the organization gone in focusing on its PM process, and engineering it to have the right balance between consistency and compliance to standards and the flexibility to enable scaling and customizing the process to the needs of individual projects, programs, or disciplines?

3. Singular methodology—To what degree has the organization integrated its PM process with its quality management, operational, financial, and other processes?

4. Benchmarking—To what degree is the organization formally seeking performance models, best practices, and feedback from other organizations and using this intelligence to improve or engineer its process?

5. Continuous improvement—To what degree is the organization evaluating its performance, capturing best practice candidates, selecting best of breed, and fine tuning the process based on performance and benchmarking results?

Within the second level, five evolutionary phases of development are identified (Figure A.1):

- Embryonic

- Executive management acceptance

- Line management acceptance

- Growth

- Maturity

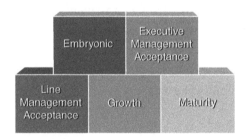

Figure A.1 Five evolutionary phases.

These phases and the rest of the model are further described below in the detailed analysis. A complete description of the model is to be found in the book *Using the Project Management Maturity Model: Strategic Planning for Project Management*, 2nd ed., by Dr. Harold Kerzner.

■ Description of the Assessment Process

The assessment process consists of the following steps:

- Initiation—Introduction of the process; selection of the assessment population

- Assessment

- Workshop discussion
- Assessment report development and delivery

A post-assessment followup should be planned to acknowledge and evaluate assessment results and plan action to improve project management performance. Maturity assessment is one part of a performance improvement program. The assessment results must be analyzed along with performance review results, current training, and methodology programs, tool set implementation, portfolio and multi-project management process improvement, and other parts of the project management improvement program. Assessment results must be explored to determine the underlying reasons for scores. The reasons posited by the KPMMM™ model are a starting point, but your specific conditions must be considered to ensure appropriate action.

■ Assessment Population

Out of the XXXX registering for the online assessment:

- XXX completed Level 1.
- XXX completed Level 2.
- XXX completed Level 3.
- XXX completed Level 4 and Level 5.

Only XXX of those completed all five levels of assessment properly.

Hence, XXX (NUMBER) assessments have been discarded from the population due to the insufficient information provided by participants.

Also, responses from groups with three or fewer participants are being excluded from the reporting but being included in the analysis.

The participants in the assessment were limited to people who were primarily from the following career development groups:

- Project managers
- project directors, sponsors, and governance personnel
- Project team members and functional managers

The above were further subdivided into the following project roles (Table A.1):

- PROJECT ROLE A
- PROJECT ROLE B
- PROJECT ROLE C
- PROJECT ROLE D
- PROJECT ROLE E
- PROJECT ROLE F
- PROJECT ROLE G
- PROJECT ROLE H
- PROJECT ROLE I

► Respondents by Project Roles and Countries Represented

Table A.1 Project Roles and Countries Represented.

Project Roles	Country 1	Country 2	Country 3	Country 4	Country 5	Country 6	Country 7	Country 8	Country 9	Country 10	Country 11	Country 12	Country 13
Project role A					1							2	
Project role B	1		11	11	11	5	18	5	32	33	8	4	15
Project role C							1	1	·4				2
Project role D						2						1	
Project role E			1	1	1		1	1	2			1	3
Project role F					1		1		3	1			
Project role G				1	1						1	1	2
Project role H					3				8		1	1	
Project role I	1	1				1	2		5		2	2	

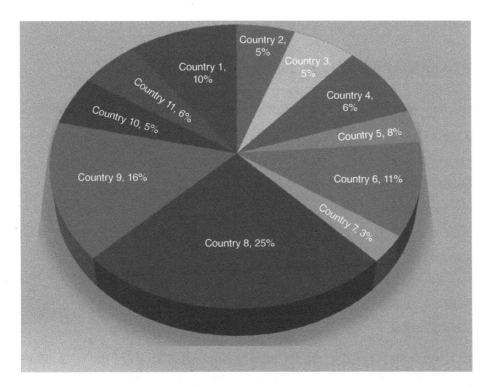

Figure A.2 XXXX Assessment by Countries Represented.

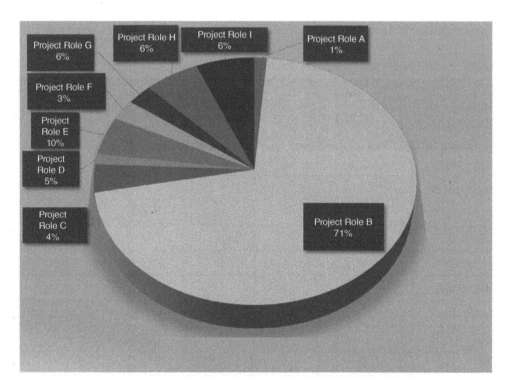

Figure A.3 XXXX Assessment by Project Roles.

▶ Executive Overview of the Assessment Results

XXXX's assessment was performed by the participation of the following business units:

- Business Unit 1
- Business Unit 2
- Business Unit 3
- Business Unit 4
- Business Unit 5
- Business Unit 6
- Business Unit 7
- Business Unit 8
- Business Unit 9

The above units were assessed for maturity at all five levels of the maturity model. This executive overview of the results is meant to provide summary information.

The detailed results are found later in the document.

The overview begins with the big picture and then provides a brief review of each level.

The Level 1 result is based on individual scores in a knowledge "exam" based on *A Guide to the Project Management Body of Knowledge* (*PMBOK® Guide*) published by the Project Management Institute (PMI).*

Levels 2–5 are based on the perceptions of respondents regarding their current work environment.

These perceptions must be validated and examined to determine an objective understanding of the organization's maturity.

Graphs show the overall scores as percentages of the maximum score for each level, showing each unit's score and comparisons with other assessed organizations.

Table A.2 shows the overall score with comparison to the model's strong and maximum scores.

Table A.2 Relative Scoring vs. Maximum Scores for Each Level.

Maturity	Your Score	Your Industry	Strong Score	Maximum Score
Level 1: Common Language	454	437	600	800
Level 2: Common Processes	18	10.33	30	60
Level 3: Singular Methodology	136	126.33	169	210
Level 4: Benchmarking	15.22	31.33	37	75
Level 5: Continuous Improvement	10.94	12.66	20	48

■ Comparison with Other Organizations

This section contains comparisons between XXXX and other organizations in the database at each maturity level.

Overall, XXXX is compared with 35 organizations that have participated in Level 1 through 5 assessments. The results show that XXXX is more mature than other organizations. This is generally an indication of an overall recognition of PM as a critical business process and a concerted effort to improve. Graphs below show the comparison results. Note that comparisons against companies of "your size" consider the size of the company and not the size of the organization within the company, where size is based on the total number of employees.

Based on the comparative analysis in Table A.2, the company looks to be more mature in some levels, especially at Level 1 and 2, with an unfavorable consistency in maturity at Levels 3, 4, and 5.

In general, there's ample room for improvements in all levels comparing to the strongest score.

The report shows that the company has been performing initiatives at the organizational level, but probably more effort should be applied in changing mindset and percolating project management knowledge through the levels.

Later in this report we'll attempt to delve into an analysis of the KPMMM assessment data, and this will hopefully help identify company strengths, weakness, and a plan for futures initiatives.

---

*PMBOK is a registered mark of the Project Management Institute, Inc.

► Level 1: Common Language (Max 800)

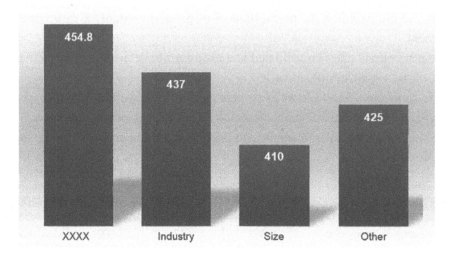

Figure A.4 Sample Level 1 Scoring.

A perfect score at this level is 800. Composite scores of 600 or greater indicate that the organization has a "very good" understanding of the project management areas of knowledge and is well positioned to begin work on Level 2 of the project management maturity model (PMMM). Scores less than 600 but greater than 300 indicate that improvement is needed but that there may be certain areas or "pockets" of formal project management. These areas may be on various levels in their knowledge of project management.

XXXX's overall score of 454.8 is below the 600 mark that is considered as being "very good" (Figure A.4). It indicates a need for project management training beyond the fundamentals level coupled with mentoring specific to the use of XXXX's project management methods within the project environment to ensure that the organization has a common project management language and that individuals have the knowledge they need.

Only the XXX respondents who completed all levels are being considered for this assessment.

Of these, 114 individuals scored between 450 and 800 at Level 1. This indicates that most individuals who participated in the assessment have a good understanding of project management common language. The breakdown of scores by delivery region for those who completed Level 1 is shown in the following sections.

In order to fully achieve a common project management language in the organization, there is a need for remediation in the form of knowledge transfer through training, coaching and mentoring, and knowledge management (for example, readily available glossaries, standard procedures, templates, just-in-time learning, real time access to best practices, white papers, and tutorials).

Analysis of individual scores for level one can be used to identify specific knowledge gaps for organization units and for individuals. To have a comprehensive training program in place, based on the comparative results, will result in positive effects, but formal classroom training is not enough. The training must be reinforced and transformed into practical applied knowledge using the approaches already stated.

Of these, initiatives for coaching and knowledge management are the most critical.

Figure A.5 shows you how your company scored within each of Level 1's subject categories. There are 10 points awarded for each correct answer. For each of these eight categories, a perfect score would be 100.

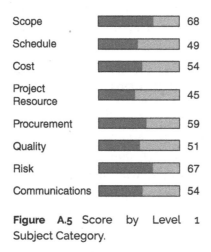

Figure A.5 Score by Level 1 Subject Category.

■ Normal Distribution and Standard Deviation Analysis

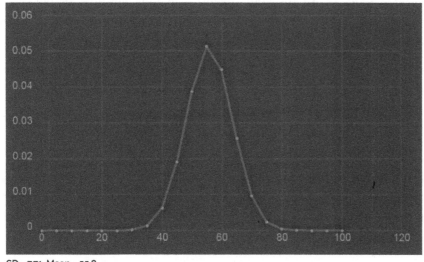

SD = 7.71, Mean = 55.8

Figure A.6 Normal Distribution and Standard Deviation Analysis.

General project management training covers the core areas of scope, schedule, cost, and risk management in the context of the project management process—initiating, planning, executing, monitoring and controlling, and closing. The importance of communication and resource management–related skills and methods should also be recognized.

The good score in scope management (68 of a possible 100) is an indication that people are aware of the importance of identifying project requirements and reaching consensus with clients, but the company needs to develop this field: creating ways to solve scope-change problems, for example, and by developing an integrated change control.

If this knowledge is translated into focused action using structured and object-oriented requirements management techniques and effective scope-change management, project performance and client satisfaction should be high.

■ Project Resource Management: Suggestions for Improvement

The lower scores in project resource management (45 of a possible 100) and schedule management (49 of a possible 100) indicate that the respondents can increase their knowledge of project estimating, scheduling, and budgeting, as well as the soft skills needed to lead and manage people, with and without authority. Education in project estimating (including scheduling and budgeting) is a first step in improving the maturity level. Other actions to improve estimating and scheduling may include improved management of estimating data, actual project results, improved time keeping, as well as coaching and mentoring of project managers in this area. Action in this area will increase the number of on-schedule and within-budget projects. Action around soft skills can reduce overall effort, avoid rework and misunderstandings, more effectively manage and avoid conflicts, better manage issues, and enable knowledge management.

An average score of 45 for your organization indicates that participants should improve their understanding of project resource management. Consider having low-scoring individuals do the following:

- Review Chapter 9, "Project Resource Management," Project Management Institute, *A Guide to the Project Management Body of Knowledge* (*PMBOK® Guide*).

- Attend courses in staff management, team building, project roles and responsibilities, conflict management, negotiating, performance appraisals, rewards, and recognition.

- Gain experience in project human resource management by getting assigned to a project team that is preparing project roles and responsibilities, preparing the organization chart, and acquiring project staff during the project initiating and planning phase.

Human resource management includes project team members, stakeholders, sponsors, customers, clients, vendors, contractors, and government employees. Leading and motivating borrowed resources is the greatest challenge to project success and often the most underestimated or overlooked by management.

■ Schedule: Suggestions for Improvement

An average score of 49 for your organization indicates that participants should improve their understanding of schedule management. Consider having low-scoring individuals do the following:

- Review Chapter 6, "Project Schedule Management," Project Management Institute, *A Guide to the Project Management Body of Knowledge* (*PMBOK® Guide*).

- Attend courses in time estimating, precedence network development, critical path, and project scheduling.

- Gain experience in project time management by getting assigned to a project team that is defining the project activities and schedule during the project initiating and planning phase.

Time is a nonrenewable resource and must be used wisely. Demands to get products to market faster do not negate the need for appropriate schedule management. Project schedule management includes defining project activities, sequencing the activities, estimating duration of activities, and developing the project schedule and control processes. Poorly defined activities and estimates lead to project failure.

■ Quality: Suggestions for Improvement

The score in quality management (51 out of a possible 100) indicates an opportunity for increased knowledge of project quality control and quality assurance. Quality control addresses activities like testing and reviews to assure that project deliverables comply with specifications. The more effectively this is done at the project level, the fewer production defects and the lower the cost of both development and ongoing operations. Quality assurance addresses the ongoing improvement of the processes used to manage and perform projects. Post project reviews and knowledge management help to improve the organization's performance over time.

An average score of 51 for your organization indicates that participants should improve their understanding of quality management. Consider having low-scoring individuals do the following:

- Review Chapter 8, "Project Quality Management," Project Management Institute, *A Guide to the Project Management Body of Knowledge* (*PMBOK® Guide*).

- Attend courses in quality planning, managing, and control.

- Gain experience in project quality management by getting assigned to a project team that is preparing a quality plan during the project initiating and planning phase.

- Participate in project reviews and audits during the project executing and controlling phases. Participate in post-project reviews and lesson-learned sessions.

Delivering an inferior product on time and within budget is disastrous to the project and can tarnish the organization's reputation. The cost of avoiding mistakes is always much less than correcting them. Therefore, quality awareness must begin during the project initiation phase and continue throughout the life of the project.

■ Communications: Suggestions for Improvement

The score in communications management (54 out of a possible 100) may indicate that the respondents can gain from increased awareness and skills training in these areas. This indicates that participants have a good understanding of communications management but could improve even more by focusing on the tools and procedures used to facilitate and manage communications as well as the documentation of accountability, roles, and responsibilities. Failure to communicate is often given as a reason for mediocre performance and failed relationships. Communications planning and implementation is often assumed as a natural consequence of working as a project team only to discover later that failed communications was the major cause of project failure. Therefore, a comprehensive communications plan, proactively implemented, is imperative for project success.

An average score of 54 for your organization indicates that participants should improve their understanding of communications management. Consider having low-scoring individuals do the following:

- Review Chapter 10, "Project Communications Management," Project Management Institute, *A Guide to the Project Management Body of Knowledge* (*PMBOK® Guide*).
- Attend courses in communications planning and control, information distribution, and performance reporting.
- Gain experience in project communications management by getting assigned to a project team that is preparing a communications plan during the project initiating and planning phase.
- Participate in performance reporting reviews during the project executing and controlling phases. Participate in project administrative closure during the project-closing phase.

Failure to communicate is often given as a reason for mediocre performance and failed relationships. Communications planning and implementation is often assumed as a natural consequence of working as a project team only to discover later that failed communications was the major cause of project failure. Therefore, a comprehensive communications plan, proactively implemented, is imperative for project success.

The scores in procurement management (59 out of a possible 100) may indicate that the respondents can gain from increased awareness and skills training in these areas. There are project environments in which the procurement of outside products and services are infrequent or limited to specific functional groups. In such organizations, low scores in this area may not indicate a need for action. However, if there are projects in which procurement is a critical factor, those project managers responsible should have knowledge of the XXXX procedures and policies and the basic principles of procurement planning, source selection, contracting, and the management of vendor-performed work under contract.

■ Procurement: Suggestions for Improvement

An average score of 59 for your organization indicates that participants should improve their understanding of procurement management. Consider having low-scoring individuals do the following:

- Review Chapter 12, "Project Procurement Management," Project Management Institute, *A Guide to the Project Management Body of Knowledge* (*PMBOK® Guide*).

- Attend courses in contract administration, contract terms and conditions, liquidated damages, performance reporting, proposal preparation and evaluation, statement of work preparation, and make or buy analysis.

- Gain experience in project procurement management by getting assigned to a project team that is preparing qualified vendor lists, open solicitations, bidders conferences, and conducting vendor evaluations during the project initiating and planning phase.

Competitive market conditions often require project managers to seek products and project assistance from outside their organization. An understanding of the solicitation process that includes proposal preparation, statements of work, bid evaluations, vendor references, and relationship management is crucial to project success.

■ Risk: Suggestions for Improvement

The composite score in risk management (53 out of a possible 100) indicates that participants have a good understanding of risk management but could improve even more. Risk management is critical to creating realistic expectations. It highlights the need to proactively identify risks, qualify them relative to their probability of occurrence, as well as quantify the impacts to the project objectives. A well-planned risk response plan will aid in the eventual monitoring and control of risk on the project. Greater demands to do more with less, partially skilled staffs, contractor workforces, geographical dispersion of project teams, time to market pressures, and rapidly changing technology add greater risk and increase the probability of project failures. The identification and monitoring of project risk is necessary for project success.

An average score of 67 for your organization, while it is not low, still indicates that participants could improve their understanding of risk management. Consider having low-scoring individuals do the following:

- Review Chapter 11, "Project Risk Management," Project Management Institute, *A Guide to the Project Management Body of Knowledge* (*PMBOK® Guide*).

- Attend courses in risk identification, quantification, and risk response planning and control.

- Gain experience in project risk management by getting assigned to a project team that is preparing a risk plan during the project initiating and planning phase.
- Participate in risk response reviews during the project executing and controlling phases.

Greater demands to do more with less, partially skilled staffs, contractor workforces, geographical dispersion of project teams, time to market pressures, and rapidly changing technology add greater risk and increase the probability of project failures. The identification and monitoring of project risk is necessary for project success.

■ Cost: Suggestions for Improvement

An average score of 54 for your organization, while it is not low, still indicates that participants should improve their understanding of cost management. Consider having low-scoring individuals do the following:

- Review Chapter 7, "Project Cost Management," Project Management Institute, *A Guide to the Project Management Body of Knowledge (PMBOK® Guide)*.
- Attend courses in work breakdown structure, bottom-up cost estimating, cost budgeting, and earned value.
- Gain experience in project cost management by getting assigned to a project team that is planning project resources, estimating costs, developing a project budget, and installing a cost control process during the project initiating and planning phase.

Estimating the project resources and the associated costs is essential to project success. Life-cycle project budgets and cost-control processes enable senior management oversight. Cost overruns are common when resources are optimistically estimated. The work breakdown structure is necessary for detailed bottom-up cost estimating.

■ Scope: Suggestions for Improvement

An average score of 68 for your organization, while it is not low, still indicates that participants could improve their understanding of scope management. Consider having low-scoring individuals do the following:

- Review Chapter 5, "Project Scope Management," Project Management Institute, *A Guide to the Project Management Body of Knowledge (PMBOK® Guide)*.
- Attend courses in scope management, defining project specifications and work breakdown structure development.
- Gain experience in project scope development by getting assigned to a project team that is developing the project scope during the project initiating and planning phase.

The scope of the project is the foundation of a project. The project scope is often developed with a work breakdown structure (WBS). Poorly developed project scope often leads to late deliverables, cost overruns, and specification disagreements. A well-developed scope with an appropriate change-control process will increase the probability of project success. Terms such as "scope creep" and "scope leap" are often associated with poorly developed project scope and no change-control process.

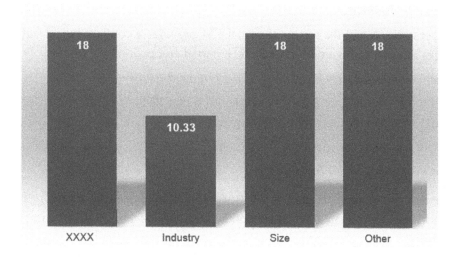

Figure A.7 Sample Level 2 Scoring.

Level 2 maturity measures the perception of respondents regarding the degree to which the organization is making a concerted effort to improve its PM performance through the implementation of standards, procedures, and methodologies.

The average score at Level 2 is 18.0 in a range from -60 +60 (Figure A.7). This score shows that XXXX is in the embryonic phase of project management maturity but is ready to move on to the second phase, which is executive management acceptance. The implications are that there is a need to make a greater effort to implement project management practices and get senior management to support the effort.

With this assessment as a baseline, XXXX will be able to see whether improvement activities are leading to greater maturity.

The figure below shows the average score within each of the Level 2 phases.

The range of possible scores for each question is between –3 (strongly disagree) and +3 (strongly agree), thus making the minimum score per Level 2 phase –12, as there are four questions per phase. Likewise, the maximum per Level 2 phase is +12. With five (5) phases and four (4) questions per phase, the aggregate range of score per individual is between –60 to +60. The score greater than 0 indicates that XXXX is heading in the right direction in terms of common processes.

Figure A.8 tells you how well your company did for Level 2 of the maturity model.

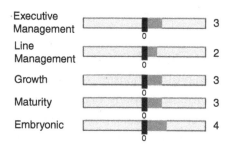

Figure A.8 Score by Level 2 Phase.

Acceptance and implementation of project management are driven by six driving forces:

- Customer expectations and the recognition that PM can help to assure consistently effective performance
- Competitiveness and the recognition that PM provides a competitive advantage and reduces internal competition to focus on external competition
- Executive understanding to promote PM from the top down
- New product development and the recognition that PM is a means to perform these projects well and to select the right projects
- Efficiency and effectiveness and the recognition that PM is a means to this end
- The need for survival of the business

There are five phases of this maturity level (called life-cycle phases). They are embryonic, executive management acceptance, line management acceptance, growth, and maturity.

It is important to note that Level 2 can and does overlap Level 1. In other words, project management training can be conducted simultaneously with the development of processes and methodologies and with the acceptance, growth, and maturity of the organization's recognition and use of PM practices.

High scores in each phase (usually +6 or greater) indicate that the corresponding phases of early maturity have been achieved. Phases with small numbers (2 or less) indicate that maturity in these areas has not yet been achieved.

Regarding the five phases at XXXX, the higher scores in the embryonic phase indicate that most BU's recognize the need for project management as well as its benefits.

■ Executive Management Acceptance: Suggestions for Improvement

Your score indicates that your organization may still not have executive management acceptance of project management. Here are some suggestions to try to gain executive management acceptance and improve your scores:

- Meet with executives and present the business case for project management and inform them what they must do to successfully implement project management, and get a commitment that they are willing to spearhead the change in the organization. Examples of visible support are executive presentations and correspondence that stress the need for project management as a core competency, executives occasionally attending project team meetings/briefings, and reviewing the ongoing progress of mission-critical projects.
- Additionally, organize and schedule executive project management sessions on the importance of their role as project sponsors and how they can accelerate the implementation of project management. Many organizations decide to provide training, develop appropriate project management tools and methodologies, and purchase state-of-the-art software and hardware in hopes that executive management will follow the lead. This approach never works and normally leads to project management

failure. Despite difficulties, executive management support and enforcement is essential to success.

Executive management acceptance of project management will not be assimilated throughout the organization without executive management acceptance, support, and enforcement. Lip service, like corporate memoranda, is not enough. Senior managers must be willing to lead the way when extensive changes are necessary in their organizations. By what they do and say, they must display a real commitment to changing the way the organization does business. Senior managers must be willing to take charge of the change process, to explain the reasons for change, and to clearly describe the new expectations they have of employees.

■ Line Management Acceptance

Your score indicates that your organization may still not have line management acceptance of project management. Here are some suggestions to try to gain line management acceptance and improve your scores:

- Meet with line managers and present the business case for project management, emphasize that there is executive support, inform them what they must do to successfully implement project management, and get a commitment that they are willing to visibly support project management in the organization. Examples of visible support are executive presentations and correspondence that stress the need for project management as a core competency, participating in project management training, occasionally attending project team meetings/briefings, reviewing the ongoing performance of their employees assigned to project teams, and releasing employees for project management training.

- Additionally, organize and schedule line manager project management sessions on the importance of their role as project stakeholders and how they can accelerate the implementation of project management.

The biggest obstacle to gaining line managers' support is a lack of executive support for project management. Few line managers would eagerly accept and support project management if they knew their superiors would not support it. Line managers do not necessarily need a strong understanding of project management tools, but they must be trained in the principles of project management. After all, the line managers are responsible for releasing employees for project management training and eventually assigning staff to project teams.

■ Growth: Suggestions for Improvement

Your score indicates that your organization may still not have achieved the growth phase of project management maturity. Here are some suggestions to try to attain and maintain the growth phase and improve your scores:

- Present to executives and line and project managers the need for and benefits of a standard methodology to cope with virtual teams and enable multiple-project management. Develop the methodology in collaboration with the stakeholders. Be careful of methodology overkill. Successful methodologies are simple and scalable to the project's needs. Integrate the methodology into the project management training. Seek senior management enforcement of the methodology. Continue to improve the methodology, emphasizing planning to minimize scope creep. When agreement is reached on how multiple projects will be managed, a software package can be selected. Unfortunately, most organizations purchase the software first and then try to figure out how to use it.

- Consider establishing a project office to proliferate project management across the entire organization.

The growth phase represents multiple-project management in an organization. This phase may coincide with the first three phases, even the embryonic phase, or run parallel to them. However, the first three phases must be complete before the growth phase can be considered complete. The management of multiple projects requires a degree of standardization not required in the previous phases and is impossible to achieve without executives, line managers, and project managers supporting and enforcing the processes. Many organizations never reach this level despite having sufficient training, appropriate project management tools, methodologies, and state-of-the-art software and hardware because they fail to get executive support and commitment to implementing project management.

■ Maturity: Suggestions for Improvement

Your score indicates that your organization may still not have achieved maturity in project management. Here are some suggestions to try to attain and maintain maturity and improve your scores:

- Present to executives, line and project managers, and the finance and accounting department the need for and benefits of a time and cost monitoring system. Integrating time and cost monitoring is not easy. It requires a total revamping of the cost-accounting system against a lot of internal resistance. However, many service companies rely on the general ledger and don't have a cost accounting system. The task will not be easy irrespective of the accounting systems situation. Executive commitment to a time and cost monitoring system is extremely important because it involves written procedures, specific training, additional software, and a commitment to thorough project planning.

- The last element of maturity is the development of a long-term educational program. Without a sustained educational program, an organization may revert to old practices very quickly. Long-term educational programs that support project management demonstrate the organization's commitment to project management. The most effective educational programs are those based on lessons learned on previous projects. In successful organizations, every project team is required to prepare such lessons-learned files. The lessons learned then are integrated in the training program.

This phase includes the development and integration of a management cost/schedule system and the development of an ongoing educational curriculum to support project management and enhance individual skills. A company must understand the importance of integrating schedule and cost management. Without such integration, no one can determine the status of a project just by looking at the schedule or the budget. Schedule and cost monitoring must become two parts of the same activity. Many companies never fully complete this life cycle because the organization is resistant to project cost control, otherwise known as horizontal accounting. Line managers dislike horizontal accounting because it clearly identifies which line managers provide good estimates for projects and which do not. Executives resist horizontal accounting because the executives want to determine a budget and schedule long before a project plan is created.

■ Embryonic: Suggestions for Improvement

Your score indicates that your organization may still be in denial about the need to improve. Here are some suggestions to try to improve your scores:

- The organization's arrogance built on past successes may be its greatest deterrent to future improvement. You may have to assess the organization to gain sufficient data to present the case to senior management.

- Recognize need. Look for symptoms of project problems, and gather facts to support the need for project management. Some examples are: schedule and cost overruns, incomplete specifications, unskilled resources, speculative technology, inadequate plans, disregard for risks, etc.

- Recognize benefits. Emphasize the benefits of project management to solve the organization's problems. Some examples are: timely delivery, reduced costs, satisfied customers, better management oversight, repeatable processes, increased profitability, etc.

- Recognize applications. Gather examples from projects to show how the application of project management tools and techniques could have avoided problems or provided early warning.

- Recognize what must be done. This is difficult because you may not know where to begin. There will be many naysayers and indifferent people in your organization unwilling to support change. Therefore, your project management improvement plan should be based on filling needs in both the short and long term. Focus on small gains that can be publicized to achieve further support. Keep everyone informed on progress.

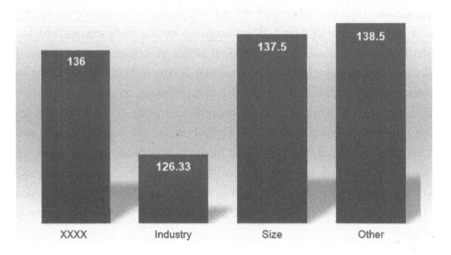

Figure A.9 Sample Level 3 Scoring.

Assessment at this level determines whether the organization recognizes the synergistic effect of combining all methodologies into a singular methodology. A significant part of this singular methodology is project management.

The synergistic approach makes process control easier than when dealing with multiple methodologies.

Singular methodology does not mean that there is a "one size fits all" approach. It means that while there may be many variations to address projects of several types, they should be unified in a common overriding process that represents the enterprise project management needs of the organization.

At 136 out of a possible 210, XXXX is at an average score (Figure A.9). A strong score at this level is 169 or above. Figure A.10 shows you how your company scored within each of Level 3's areas.

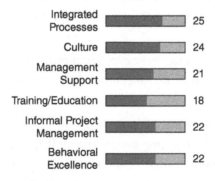

Figure A.10 Score by Level 3 Area.

■ Integrated Processes: Suggestions for Improvement

Your score indicates that your organization has not realized the importance of a single integrated methodology. Here are some suggestions to try to improve your scores:

- Your organization must recognize that multiple processes can be streamlined into one integrated process that encompasses all other processes.

- Present to executives and line and project managers the need for and benefits of a single integrated process.

- Stress the avoidance of duplication of effort, resources, and facilities, and the reduction in training costs.

- Develop the integrated process in collaboration with the stakeholders. The processes must be integrated, not bundled; therefore, there may be resistance from process owners as to what is added or left out of the integrated process. Be careful of process overkill, and develop a process that is scalable to the project needs.

- Pilot test the process.

- Continue to improve the process.

- Incorporate lessons learned. Creating a singular integrated process that encompasses all other processes leads to organizational efficiency and effectiveness.

In the past decade, many companies have developed and implemented processes (Quality Management, Project Management, Change Management, Systems Life Cycle Management, Concurrent Engineering, Risk Management) with the intent to solve problems or improve effectiveness. Competing departments, which often got results contrary to the original intent, developed these processes. These processes competed for precious resources, appeared to be duplicative, caused confusion, and quickly lost management support within the organization, resulting in processes that were seldom used. Companies that are relatively immature in project management have multiple processes in place. As companies recognize the synergistic effects of combining management processes into one system, usually the first two processes integrated are project management and total quality management.

■ Culture: Suggestions for Improvement

Your score indicates that your organization's culture does not support project management. Here are some suggestions to try to improve your scores:

- Project management can't be copied from one company to another. Each organization is unique. But benchmarking best practices in leadership, management, and operations against competing organizations may be helpful.

- The organization must support the four basic values of project management: cooperation, teamwork, trust, and effective communications.

- Accountability for project success is shared equally between the project manager and the line manager.

- Rewards for project success are given to the project team, not restricted to politically aligned employees.

- The company must consider the development of a single project management process that clearly defines the project manager's role and is supported by senior management.

Corporate cultures can't be changed overnight. They change over years. The successful implementation of project management usually requires cultural change. Excellence in project management is achieved when the culture of the company can change quickly to handle the demands of new and multiple projects. Yet the culture must be equipped to adapt to constantly evolving and dynamic business environment. Excellent companies cope with change in real time and live with the potential chaos that comes with it.

■ Management Support: Suggestions for Improvement

Your score indicates that your organization has not attained the level of management support needed for a successful integrated singular methodology. Here are some suggestions to try to improve your scores:

- Present to senior management the importance of their support and enforcement to make the singular methodology work.
- Stress visible support and involvement, not just lip service.
- Roles and responsibilities of project sponsors, project managers, and line managers may need development or focus.
- Review the project charter guidance to determine the extent of the project manager's level of empowerment for decision-making and establishing relationships with sponsors, customers, and line managers.

Senior managers are the architects of corporate culture. They are charged with making sure their company's culture, once accepted, doesn't come apart. Visible management support is essential to maintaining a project management culture. Project management permeates the organization throughout all layers of management. The support is visible. Each layer or level of management understands their role and the support needed to make a singular methodology work.

■ Training and Education: Suggestions for Improvement

Your score indicates that your organization does not have a sufficient project management training and education program. Here are some suggestions to try to improve your scores:

- Survey the organization to determine the training and education needs. Review the internal project management curriculum.
- Determine whether there is a balance between quantitative and behavioral courses. Ensure that the integrated singular methodology is taught in all project management courses.
- Ensure that the training is available to all team members, not just project managers.
- Provide tailored courses to all levels of employees.
- Stress the integration of lesson learned in courses.
- Stress the inclusion of project management certification courses in the curriculum.

Effective project management training and education must be balanced between quantitative and behavioral courses and tailored to all employee levels. Project management training and education is an investment, and, as such, senior management wishes to know when the added profits will materialize. Initially, there may be a substantial cost incurred. But as the culture develops and informal project management matures, the cost of project management diminishes to a stable level while the additional profits grow.

■ Informal Project Management: Suggestions for Improvement

Your score indicates that your organization doesn't have the level of trust in project managers to execute informal project management. Here are some suggestions to try to improve your scores:

- Conduct a survey in your organization to determine if the project management methodology is too rigid and solicit suggestions for improvements.

- Where possible, lessen the project documentation, reporting requirements, and time spent in project review meetings.

- Try to eliminate redundant meetings.

- Review the scalability of the methodology, and try to reduce mandatory requirements that belie trust in the project managers.

- Determine if managers and employees freely share information or if there is a "shoot the messenger" mentality present.

With management support and a corporate culture of trust, the integrated singular methodology becomes more of a framework than rigid policies and procedures. Paperwork is minimized. For this to work effectively, there must be trained and empowered project managers, effective communications, cooperation, trust, and teamwork. Informal project management drastically lowers methodology execution cost and execution time.

■ Behavioral Excellence: Suggestions for Improvement

Your score indicates that your organization has not realized the importance of behavioral skills. Here are some suggestions to try to improve your scores:

- Present to senior management the importance of behavioral skills necessary to make the integrated singular methodology work.

- Develop an atmosphere conducive to good working relationships among the project manager, parent organization, and client organization.

- Encourage openness and honesty within the organization.

- Stress delegation of sufficient authority to project managers.

- Create an atmosphere of cooperation and collaboration.

- Stress managerial, operational, and product integrity.

- Recognize and reward project team members for their performance.

- Encourage initiative, creativity, and innovation.

- Maintain an open-door policy.
- Encourage risk taking, and allow employees to make mistakes.

► Level 4: Benchmarking (Max 75)

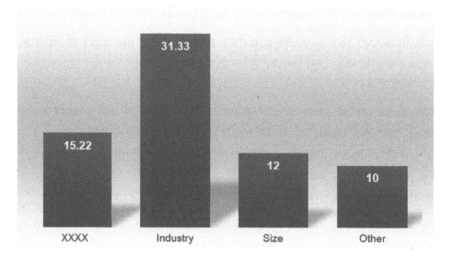

Figure A.11 Sample Level 4 Scoring.

The average score at Level 4 is 15.2 of a possible 75 (Figure A.11), with a strong score being above 37, pointing out that while there is some benchmarking being done it is not institutionalized and consistently used. See the detailed analysis for a further breakdown of this level. The level of maturity of benchmarking is a foundation for continuous improvement and stimulates maturity at all levels.

Figure A.12 tells you how well your company did for Level 4 of the maturity model.

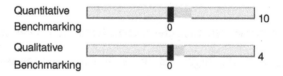

Figure A.12 Level 4 (Benchmarking) Scores.

■ Quantitative Benchmarking: Suggestions for Improvement

Your score indicates that your organization has not realized the importance of quantitative benchmarking. Here are some suggestions to try to improve your scores:

- Your company must be committed to project management across the entire organization.
- Quantitative benchmarking analyzes processes and methodologies. To accomplish quantitative benchmarking, your organization must establish a project management office (PMO) or center of excellence (COE) for project management. The project management office or COE must be dedicated to the project management

improvement process. The project office can improve the enforcement and compliance of a singular methodology, tighter cost controls, efficient resource allocations and utilizations, quality, scope, and risk management.

- Benchmarking must be made against both similar and nonsimilar industries.
- Continuous improvement is best accomplished through continuous benchmarking.

Quantitative benchmarking in organizations is the systematic comparison of elements of performance in an organization against those of other organizations, usually with the aim of mutual improvement. Benchmarking should not be performed unless the organization is willing to make changes.

The changes must be part of the structured process that includes evaluation, applicability, and risks. Benchmarking is part of the strategic planning process for project management that results in an action plan ready for implementation. Project management benchmarking is the process of continuously comparing the project management practices of your organization with leaders anywhere in the world to gain information to help you improve your performance.

The information can be used to help you improve your processes and how they are executed, or the information can be used to help your company become more competitive in the marketplace. Project management benchmarking addresses quantitative process improvement opportunities and qualitative process improvement opportunities.

■ Qualitative Benchmarking: Suggestions for Improvement

Your score indicates that your organization has not realized the importance of qualitative benchmarking. Here are some suggestions to try to improve your scores:

- Your company must be committed to project management across the entire organization.
- Qualitative benchmarking looks at project management applications. To accomplish qualitative benchmarking, your organization must establish a project office (PO) or center of excellence (COE) for project management. The project office must obtain corporate acceptance of a singular methodology for managing projects.
- Increase the usage and support of existing users, and attract new internal users.
- Discourage the development of parallel methodologies by emphasizing the present and future benefits of using a singular methodology.
- Integrate other company processes into a singular methodology. Provide software enhancements to support integrated processes and the singular methodology.
- Benchmarking must be made against both similar and nonsimilar industries. Continuous improvement is best accomplished through continuous benchmarking.

Qualitative benchmarking in organizations is the systematic comparison of elements of performance in an organization against those of other organizations, usually with the aim of mutual improvement. Benchmarking should not be performed unless the organization is willing to make changes. The changes must be part of the structured process that includes evaluation, applicability, and risks. Benchmarking is part of the strategic planning process for project management that results in an action plan ready for implementation. Project management benchmarking is the process of continuously comparing the project management practices of your organization with leaders anywhere in the world to gain information to help you improve your performance.

The information can be used to help you improve your processes and how they are executed, or the information can be used to help your company become more competitive in the marketplace. Project management benchmarking addresses quantitative process improvement opportunities and qualitative process improvement opportunities.

▶ Level 5: Continuous Improvement (Max 48)

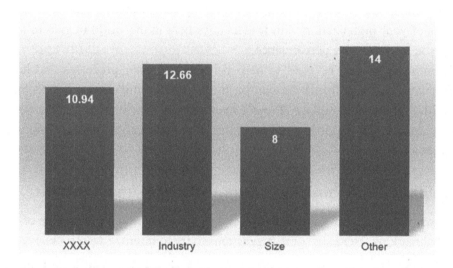

Figure A.13 Sample Level 5 Scoring.

The average for Level 5 is 10.94 of a possible 48 (Figure A.13), with strong being above 20. This indicates that XXXX can gain by cultivating a continuous improvement culture supported by a knowledge management process that includes formal lessons-learned processes, performance measurement and analysis, problem identification, and cause-and-effect analysis followed by methodology updates and other actions. While Level 5 is the highest level, growth in this area promotes improvement in all prior levels and is directly related to benchmarking.

Continuous improvement requires an environment that eliminates blaming and replaces it with an attitude of learning. As the organization learns from its issues and shortfalls, it improves its processes and, through them, its performance.

Figure A.14 tells you how well your company did for Level 5 of the continuous improvement model.

Figure A.14 Level 5 (Continuous Improvement) Score.

■ Continuous Improvement: Suggestions for Improvement

Your score indicates that there is a strong resistance to change in your organization and a lack of senior management support for continuous improvement. Success makes it difficult to implement changes if people must be removed from their comfort zones. You may have to wait until there is pressure from your customers or an erosion of your business base before you get senior management attention. Here are some suggestions to try to improve your scores:

- A learning organization capitalizes on its mistakes. You need to create a lessons-learned file that can be accessed by managers. Lessons learned must be used to update methodology, forms, training curriculum, and the mentor program. Review your methodology to ensure that it is still in sync with the corporate strategy, then adjust accordingly.

- Consider adding additional functions and staff to the project office to implement continuous improvement.

- Continue to communicate the project management successes and the importance of the project office. Consider an assessment by an outside consultant.

The organization evaluates the information learned during benchmarking and implements the changes necessary to improve the project management process. It is in this level that the company comes to the realization that excellence in project management is a never-ending journey. The benefits of continuous improvement include better competitive positioning, corporate unity, improved cost analysis, customer value added, better customer expectations, and ease of implementation. If an organization doesn't continue to improve, it will soon fall behind its competitors.

▶ A Study of Level 1 Performance

■ Level 1: Geographical Comparisons by Region

Figures A.15 and A.16 compare results by geographical regions.* The implications are that there is a need to bring up the current levels of maturity in the various geographic regions that XXXX operates in and to analyze the regional differences at other levels to better understand the reasons for them and to determine how to leverage the best practices and advanced maturity in some regions. The P region appears to be the strongest in terms of having a common project management knowledge. There also seems to exist a strong deviation from the mean in terms of the performance of the different regions.

---

* The graphs in the following sections were taken from an actual study. For privacy, any potentially identifiable information has been removed, including labels on some charts that do not affect their utility for these examples.

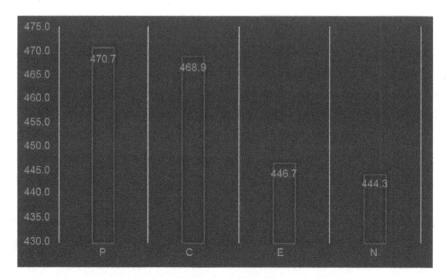

Figure A.15   Geographical comparisons by region.

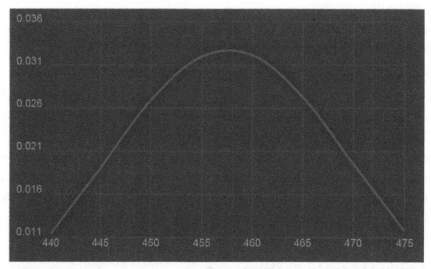

S.D = 12.1, Mean = 457.6

Figure A.16   Normal distribution and standard deviation analysis.

■ Level 1: Geographical Comparisons by Country

Figures A.17 and A.18 compare results by country. The implications are that there is a need to bring up the current levels of maturity in the various countries that XXXX operates in and to analyze the regional differences at other levels to better understand the reasons for them and to determine how to leverage the best practices and advanced maturity in some regions. Country 7 leads the assessment at Level 1, indicating a strong common language of project management prevalent in the offices situated in that country. The second-strongest country seems to be Country 12, and other countries, all the way to Country 6, seem to be fairly close in terms of relative performance to each other.

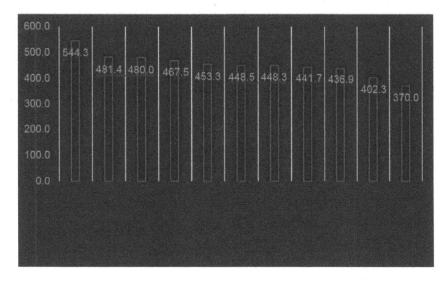

Figure A.17   Geographical comparisons by country.

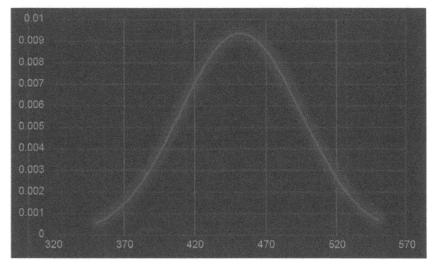

S.D. = 42.6, Mean = 452.2

Figure A.18   Normal distribution and standard deviation analysis.

■ Level 1: Comparisons by Project Roles

Figures A.19 and A.20 depict the average scores by project roles.

Purchasing and contracts, by being involved in detailed project management with legal ramifications, seem to have the strongest maturity at this level. There are some issues with lack of oversight of project management.

The primary differences could be seen to be the application of PM processes to daily work. Further analysis is needed to pinpoint the reasons for the differences and then address them accordingly.

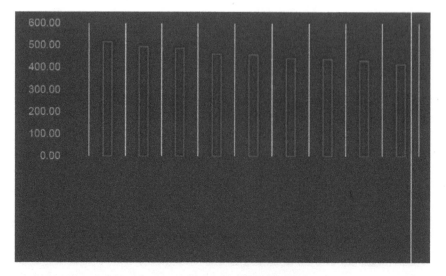

Figure A.19 Comparison of scores between project roles.

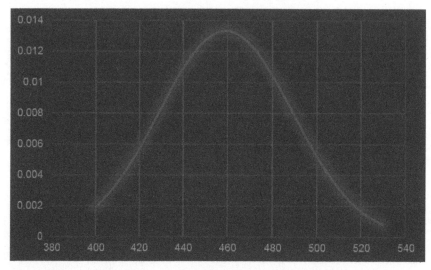

S.D. = 29.9, Mean = 458.9

Figure A.20 Normal distribution and standard deviation analysis.

■ Level 1: Comparisons by Career Developmental Groups

Figures A.21 and A.22 depict the average scores by career development groups.

Project sponsors lead the group, and it is evident that their understanding of project management is strongest by their work experience in implementing and managing projects. Further assessment is required to bring the lagging career development groups up to the company standard.

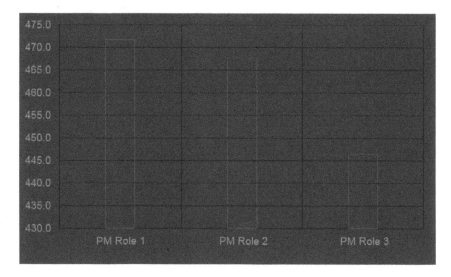

Figure A.21 Comparison of scores between career development groups.

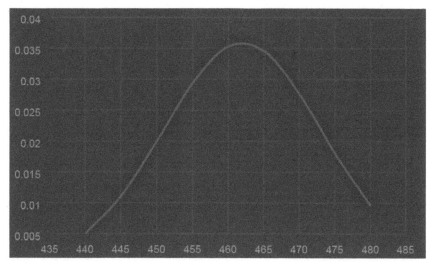

S.D. = 11.1, Mean = 461.9

Figure A.22 Normal distribution and standard deviation analysis.

▶ A Study of Level 2 Performance

■ Level 2: Geographical Comparisons by Region

Figures A.23 and A.24 compare results by geographical regions. The implications are that there is a need to bring up the current levels of maturity in the various geographic regions that XXXX operates in and to analyze the regional differences at other levels to better understand the reasons for them.

The C region leads the assessment at this level, showing a higher level of understanding of the need for common processes. This may also indicate that a consistent set of processes are being followed in the C region.

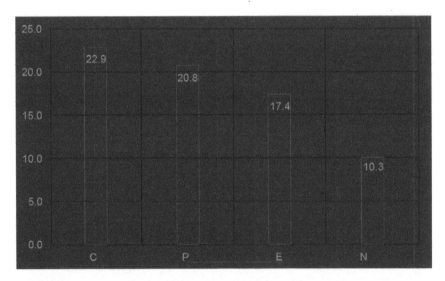

Figure A.23   Geographical comparisons by region.

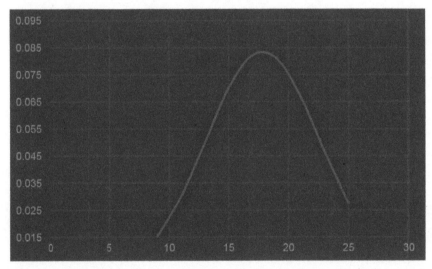

S.D. = 4.7, Mean = 17.8

Figure A.24   Normal distribution and standard deviation analysis.

■ Level 2: Geographical Comparisons by Country

Figures A.24 and A.25 compares results by country. The implications are that there is a need to bring up the current levels of maturity in the various countries that XXXX operates in and to analyze the regional differences at other levels to better understand the reasons for them and to determine how to leverage the best practices and advanced maturity in some regions. Country 3 tops this level of maturity in understanding the importance of common processes in building the maturity of an organization. The countries on the tail end seem to have a rather different perspective from those at the top.

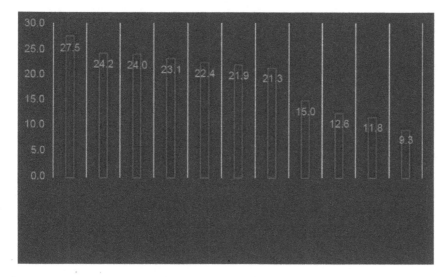

Figure A.25 Geographical comparisons by country.

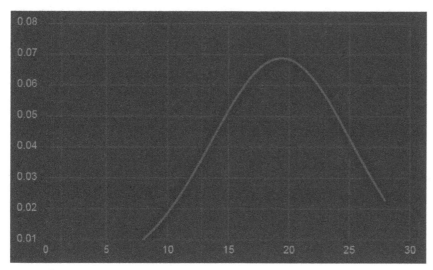

S.D. = 5.7, Mean = 19.3

Figure A.26 Normal distribution and standard deviation analysis.

■ Level 2: Comparisons by Project Roles

Figures A.27 and A.28 depict the average scores by project roles. Safety and environment, among all the project roles within XXXX, understand the need for common processes best, and it is understandable given the predominant use of standardized processes for safety.

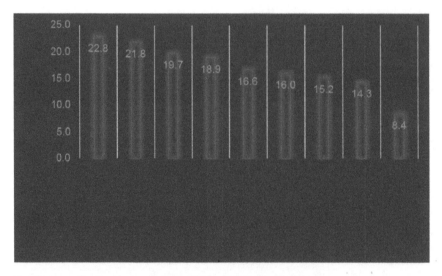

Figure A.27 Comparison of scores between project roles.

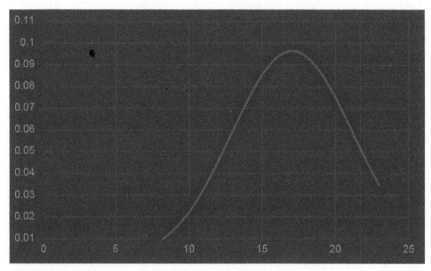

S.D. = 4.17, Mean = 17.06

Figure A.28 Normal distribution and standard deviation analysis.

■ Level 2: Comparisons by Career Developmental Groups

Figures A.29 and A.30 depict the average scores by career development groups. The integral development career group leads the assessment at this level, and from this it is evident that they possess the maturity to understand the need for common processes best among other developmental groups. Further assessment is required to bring the lagging career development groups up to the company standard.

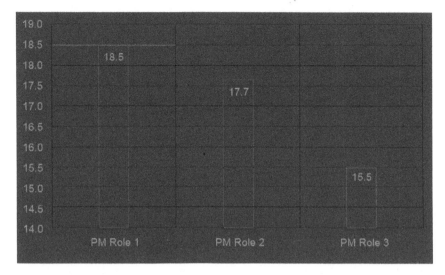

Figure A.29 Comparison of scores between career development groups.

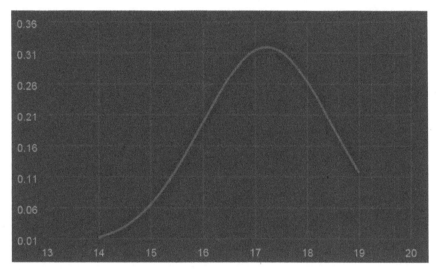

S.D.= 1.24, Mean = 17.22

Figure A.30 Normal distribution and standard deviation analysis.

▶ A Study of Level 3 Performance

■ Level 3: Geographical Comparisons by Region

Figures A.31 and A.32 compare results by geographical regions. The implications are that there is a need to bring up the current levels of maturity in the various geographic regions that XXXX operates in and to analyze the regional differences at other levels to better understand the reasons for them and to determine how to leverage the best practices and advanced maturity in some regions. The P and C regions lead the assessment at this level, showing a higher level of understanding of the need for a singular methodology. From the Level 2 scores, region-wise, it is logical that C also has a higher maturity in this level, as having common processes usually implies the use of a singular methodology.

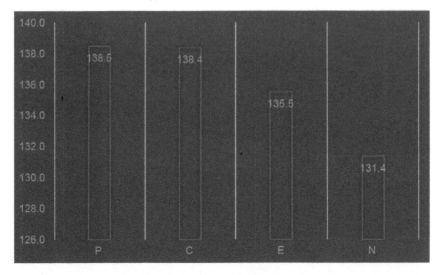

Figure A.31 Geographical comparisons by region.

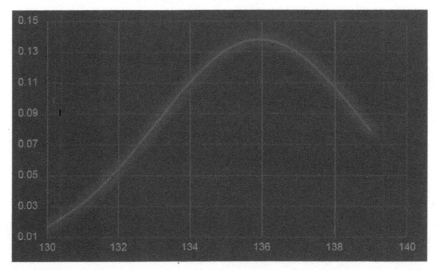

S.D. = 2.8, Mean = 135.9

Figure A.32 Normal distribution and standard deviation analysis.

■ Level 3: Geographical Comparisons by Country

Figures A.33 and A.34 compares result by country. Country 12 and Country 3 top this level of maturity in understanding the importance of having a singular methodology. The countries on the tail end (Country 4, Country 8, Country 9, Country 6, Country 10, and Country 5) seem to have a largely different perspective from those at the top. The countries with scores at the bottom also have lower deviation, i.e., they are consistent in their levels of understanding/maturity.

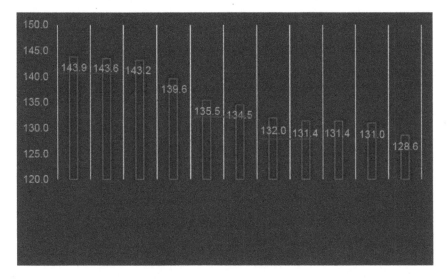

Figure A.33 Geographical comparisons by country.

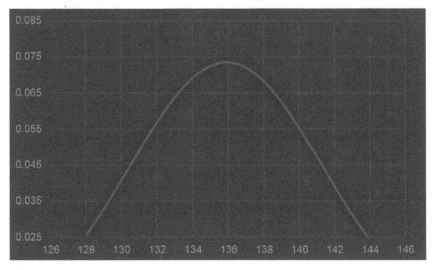

S.D. = 5.4, Mean = 135.8

Figure A.34 Normal distribution and standard deviation analysis.

■ Level 3: Comparisons by Project Roles

Figures A.35 and A.36 depict the average scores by project roles. Project role A, among all the project roles within XXXX, understands the need for a singular methodology best, and it is understandable given that project role A is typically a department of strategic advantage within XXXX and quite possible extensively uses a singular methodology. Further assessment is required to determine the rationale behind the scoring at both the top and the tail end.

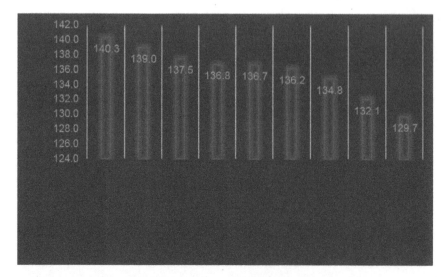

Figure A.35 Comparison of scores between project roles.

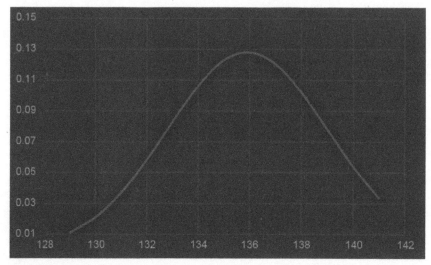

S.D. = 3.1, Mean = 135.9

Figure A.36 Normal distribution and standard deviation analysis.

■ Level 3: Comparisons by Career Developmental Groups

Figures A.37 and A.38 depict the average scores by career development groups.

The project sponsors career group leads the assessment at this level, and from this it is evident that they possess the maturity to understand the need for a singular methodology best among other developmental groups. It is believed that this comes from their extensive experience directing and managing projects with a consistent, singular methodology. Further assessment is required to bring the lagging career development groups up to the company standard.

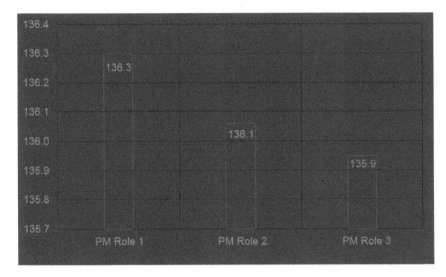

Figure A.37   Comparison of scores between career development groups.

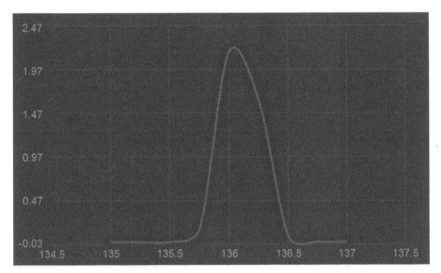

S.D. = 0.14, Mean = 136.1

Figure A.38   Normal distribution and standard deviation analysis.

► A Study of Level 4 Performance

■ Level 4: Geographical Comparisons by Region

Figures A.39 and A.40 compare results by geographical regions. The P region leads the assessment at this level, showing a higher level of understanding of the need for benchmarking and comparing its relative performance to other similar businesses.

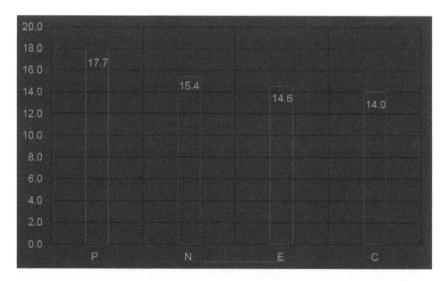

Figure A.39 Geographical comparisons by region.

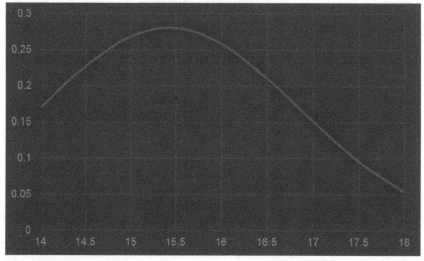

S.D. = 1.4, Mean = 15.4

Figure A.40 Normal distribution and standard deviation analysis.

■ Level 4: Geographical Comparisons by Country

Figures A.41 and A.42 compare results by country. Country 8 tops this level of maturity in understanding the importance of benchmarking with similar industries. A logical explanation for the performance of the country in this maturity level may be the above-average presence of XXX industries in this country. The country at the tail end, Country 6, seems to have a rather different perspective from those at the top, with a significant deviation from the main scores.

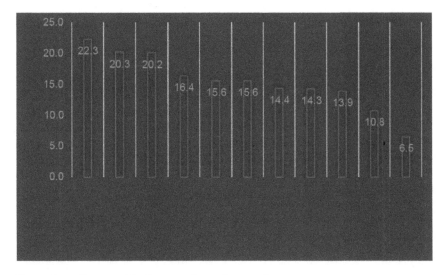

Figure A.41 Geographical comparisons by country.

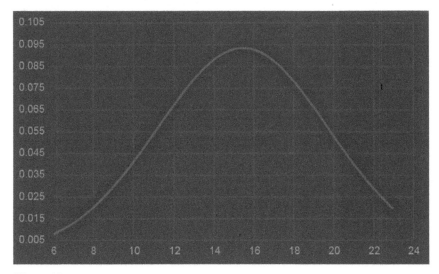

S.D. = 4.2, Mean = 15.4

Figure A.42 Normal distribution and standard deviation analysis.

■ Level 4: Comparisons by Project Roles

Figures A.43 and A.44 depict the average scores by project roles.

Project role A, among all the project roles within XXXX, again seems to understand the need for benchmarking best, and it is understandable given that project role A scored highest in understanding the need for a singular methodology and quite possibly extensively benchmarks to keep this methodology updated. Further assessment is required to determine the rationale behind the scoring at both the top and the tail end to draw from what is being done right and wrong.

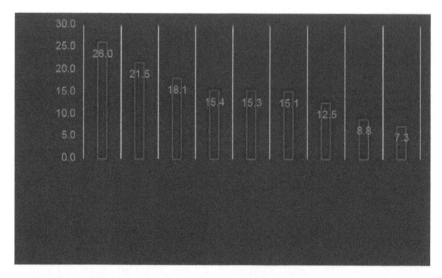

Figure A.43   Comparison of scores between project roles.

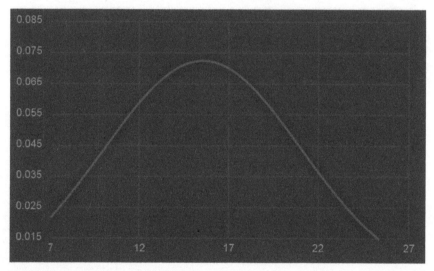

S.D. = 5.5, Mean = 15.5

Figure A.44   Normal distribution and standard deviation analysis.

■ Level 4: Comparisons by Career Developmental Groups

Figures A.45 and A.46 depict the average scores by career development groups.

The integral development career group leads the assessment at this level, and from this it is evident that they possess the maturity to understand the need for benchmarking among other industries. Further assessment is required to bring the lagging career development groups up to the company standard, especially the project team members and functional managers.

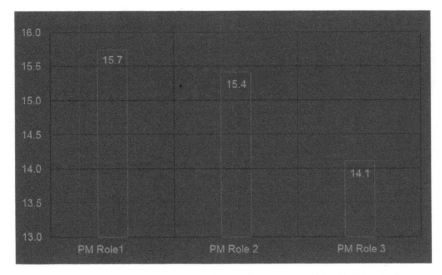

Figure A.45 Comparison of scores between career development groups.

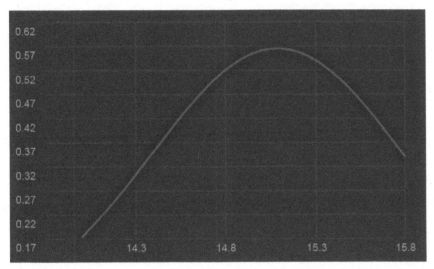

S.D. = 0.7, Mean = 15.0

Figure A.46 Normal distribution and standard deviation analysis.

▶ A Study of Level 5 Performance

■ Level 5: Geographical Comparisons by Region

Figures A.47 and A.48 compare results by geographical regions.

The P region leads the assessment at this level too, showing a higher level of understanding of the need for continuous improvement. This is synonymous with the P leadership at level 4, as any attempt to show maturity in benchmarking would implicitly involve continuous improvement as well.

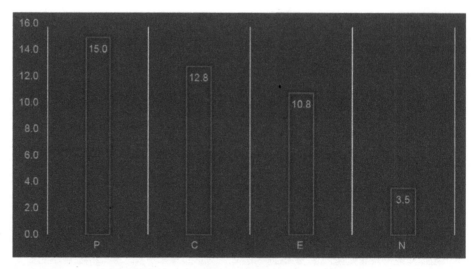

Figure A.47 Geographical comparisons by region.

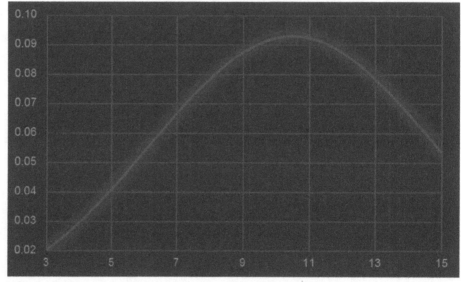

S.D. = 4.2, Mean = 10.5

Figure A.48 Normal distribution and standard deviation analysis.

■ Level 5: Geographical Comparisons by Country

Figures A.49 and A.50 compare results by country.

Country 8 tops this level of maturity in understanding the importance of continuous improvement as it did for the benchmarking maturity level, as both concepts usually complement and follow each other.

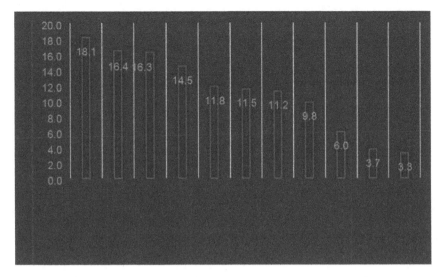

Figure A.49 Geographical comparisons by country.

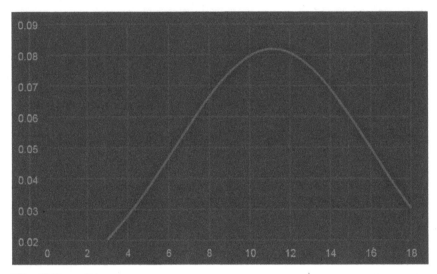

S.D. = 4.8, Mean = 11.1

Figure A.50 Normal distribution and standard deviation analysis.

■ Level 5: Comparisons by Project Roles

Figures A.51 and A.52 depicts the average scores by project roles.

Project role I, among all the project roles within XXXX, seems to understand the need for continuous improvement best. Further assessment is required to determine the rationale behind the scoring at both the top and the tail end to draw from what is being done right and wrong, especially with the finding that project role A, a strong contender at the benchmarking maturity level, scored at the tail end in this assessment.

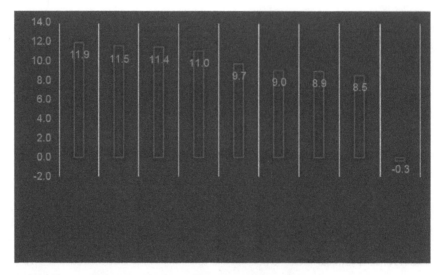

Figure A.51 Comparison of scores between project roles.

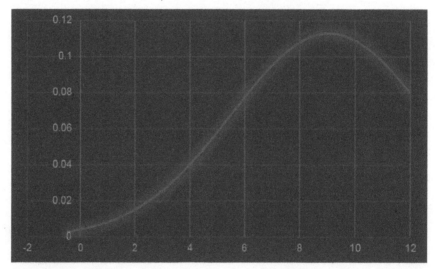

S.D. = 3.5, Mean = 9.1

Figure A.52 Normal distribution and standard deviation analysis.

■ Level 5: Comparisons by Career Developmental Groups

Figures A.53 and A.54 depict the average scores by career development groups.

The project managers career group leads the assessment at this level, and from this it is evident that they possess the maturity to understand the need for continuous improvement. Further assessment is required to bring the lagging career development groups up to the company standard, especially the project team members and functional managers.

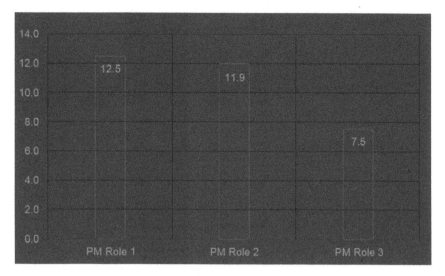

Figure A.53 Comparison of scores between career development groups.

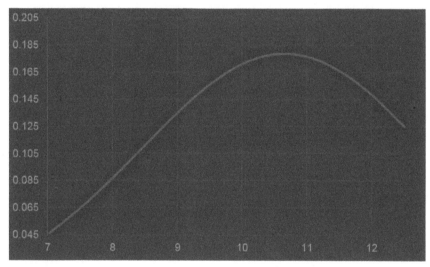

S.D. = 2.2, Mean = 10.6

Figure A.54 Normal distribution and standard deviation analysis.

▶ Suggested Actions

Overall, the assessment results should be analyzed among a small group of key PM champions and subject matter experts to identify open questions, reaffirm goals and objectives, and plan next steps in the form of an improvement program plan.

Cause analysis to explore deficiency causes is an essential next step. For example, the assessment indicates the need for further improvement in cultivating a common language and knowledge base. Whether that should be addressed by formal education, more integrated just-in-time knowledge transfer, community practice events, better articulated and delivered procedures and templates, or other means can only be determined through root cause analysis followed by planning to address the causes.

Index